The Hidden Power of Smell

Paul A. Moore

The Hidden Power of Smell

How Chemicals Influence Our Lives and Behavior

Paul A. Moore
Laboratory for Sensory Ecology
Bowling Green State University
Bowling Green, OH, USA

ISBN 978-3-319-15650-7 ISBN 978-3-319-15651-4 (eBook)
DOI 10.1007/978-3-319-15651-4

Library of Congress Control Number: 2015945950

Springer Cham Heidelberg New York Dordrecht London

Printed on acid-free paper

Springer International Publishing AG Switzerland is part of Springer Science+Business Media (www.springer.com)

To Jen for showing me love,
To Meghan for showing me empathy,
To Connor for showing me humor,
and
To Nobody, because even Nobodies
are somebodies.

Preface

This book provides a mere glimpse into the world that has dominated my career for more than three decades. In studying the minds of animals and attempting to elucidate the role that chemical signals play in decision making, one gets quite close to the subject matter and the organisms involved. At times, this myopic view of nature and its organisms hinders the ability of someone to see the bigger picture of how everything fits together into a coherent picture. Given the structure of science today and the need to become an expert in one area, we are forced to focus and specialize. Writing this book has forced me to read outside of my focal point (aquatic chemical ecology) and engage my mind with researchers and animals far afield from my crayfish. In doing this mental expansion, I found myself amazed and excited about all of the ways that evolutionary processes have developed adaptations that involve the use of the chemical world. Plants that communicate fear signals and animals that cross-dress using special perfumes are just a couple of other worldly adaptations that are often overlooked. I hope that this renewed excitement found its way into this book and I capture the imagination of the reader.

One of the most common questions asked of me is how I got to this point in time. How did I make the decision to study marine and aquatic life? Why lobsters and crayfish? Why chemical signals? I enjoy these questions because they provide me an opportunity to pause and think about seminal events that caused me to choose one path over another. In retrospect, many of these paths look quite linear, but in reality, they are most likely a life-sized game of pinball. Sometimes, we think we know where we are headed and out of the blue a bumper comes along and redirects our life.

As a youth, I would spend my Saturdays watching one of three programs: Saturday morning cartoons, B-grade monster movies like Godzilla or Mothra, or *The Undersea World of Jacques Cousteau*. More than likely, this latter show is the one which captured my imagination about the aquatic world that started me down a path that has taken me to my career. The images shown by the weekly adventures of Cousteau and his team could have been of an alien world. Deep sea corals, sea cucumbers, and large cephalopods were shown in such detail that I imagined that I was really underwater. Because this was before the internet allowed one to find an image and description of some deep sea creature instantly, I really felt that I was privileged to be a part of the discovery of new and fascinating organisms. Cousteau would focus on the diversity of life instead of running a popular culture-driven theme week that merely centered on the charismatic megafauna of sharks and dolphins.

Looking backward on one's life can produce a false sense of predestined steps that ultimately lead to a singular consequence and that consequence is where we are in the present. Probably in reality, a series of innocuous or random events have led us to where we end up. The TV show on the marine environment was a constant narrative of my youth that played against summer vacations spent really exploring the natural world. Every summer, at least for a week if not longer, we would pack up the family in a camper and explore the less charted areas of the United States. This first started in my home state of Michigan with yearly trips to the Upper Peninsula and its many bodies of water and rivers. I fondly remember playing in rivers, looking for crayfish (which would be foreshadowing if our lives were movies), and using rocks, twigs, and branches to redirect the water. I was becoming fascinated with the movement of water through, around, and over the many different obstacles in the northern Michigan streams and rivers. These two factors, Jacque Cousteau and the aquatic life of Michigan, interacted to create inside of me a draw to water in some form.

Maybe the final event that turned into destiny was a simple little job report written around the age of 10 while I was in fourth grade. Through some chance or guided by the unconscious images of undersea life, I selected to do a report on Biological Oceanography. Probably not the most typical job report for most kids in the fourth grade, but the details that I wrote about cemented in my mind that this was the greatest job in the world. Thus, every academic decision I made from that point forward was aimed at achieving that singular goal. I was going to become a biological oceanographer.

These series of events began my love affair with the aquatic life and on a pathway that went from undergraduate work, to graduate work, and on to academic work in the field of science. Being somewhat adventurous, the academic training has allowed me to travel to and explore different aquatic habitats. Working in submersibles, walking on salt flats, diving on crystal clear reefs, slogging through muddy bogs, to some of the most beautiful streams and rivers on the plant, I found one commonality across all of these disparate habitats. Aquatic life is literally awash with chemicals, and organisms have evolved an incredible ability to harness these signals to carry out the basic functions of life. The more that I (and all of the other scientists working in these fields) answered one set of questions, a dozen more popped up. Rather than being frustrated with this increasing level of planned ignorance, I become more and more excited. Science is one of those endeavors where the more you know, the more you recognize that is left to know.

The aquatic realm, in particular, is a habitat that is often devoid of light, so visual signals are not very functional. So, organisms have evolved elaborate mechanisms to use chemical signals to carry out all of the functions that terrestrial life uses visual and auditory cues. The chemical sense has been termed the primal sense because it is the first sensory system to evolve and as such, is deeply connected to evolutionary history that all organisms share. A description of any environment, especially aquatic ones, would be incomplete without highlighting odors.

So, this book is born out of the love affair between aquatic life, chemical signals, and the need to share what I know about this fabulous field. I have covered only a small fraction of the stories and science of chemical ecology. As touched on at the end of each chapter and expounded on in the last chapter, human chemical ecology is still poorly undeveloped compared to our understanding of other animal's use of chemical signals, and that understanding is sadly decades behind the fields of vision and hearing. I am by no means an expert on human chemical ecology, but I offer up these stories in order to enlighten our collective understanding on this hidden sense.

Bowling Green, OH, USA Paul A. Moore

Acknowledgements

To stand on the shoulders of giants is a paraphrase of Sir Issac Newton's famous quote. If we are honest as scientists, we all stand on the shoulders of greater giants and lesser giants. Our thinking and work would not be possible if there were no researchers that opened the doors of knowledge in front of us and gladly shared their work. This book, although my words, is a synthesis of countless numbers of students, faculty, researchers, and artists that have laid the foundation of knowledge on which I stand. Some of those researchers are named in this book whereas others are hidden just like the chemical signals that they study. I want to recognize all of those individuals (both named and unnamed) for their hard work in trying to bring to light nature's secrets. Nature often guards those secrets well from the prying eyes, ears, and noses of scientists.

The ideas in this book have arisen from countless conversations over dinner or alcohol with friends and colleagues. These include dear friends like Rainer Voigt and Carl Merrill and advisor Jelle Atema who were there at the beginning of my scientific career. These individuals play an instrumental role in launching my early career path. I have been privileged to have spent considerable periods of intellectual growth in some of the most groundbreaking environments that a young scientist could every want. I grew up scientifically in Woods Hole Massachusetts at the Marine Biological Laboratory which itself is surrounded by three other scientific institutions (Woods Hole Oceanographic Institute, National Marine Fisheries, and the United States Geological Survey). I interacted with undergraduates, graduate students, scientists, and Nobel Laureates. If one wants to do Marine Biology, there are few better environments. As a postdoctoral researcher, I spent 2 years at the Monell Chemical Senses Center surrounded by leading researchers all focused on the smell, taste, and trigeminal senses. My office mates, particularly Paul Breslin, my advisor, Bruce Bryant, and two rouge scientists, Larry Clark and Russ Mason, played an important role in cementing my knowledge of the chemical senses. Today, I am lucky enough to spend my summers at the University of Michigan Biological Station where some of the leading ecological researchers gather each summer to teach and perform research. These researchers including Rex Lowe, Steve

Pruett-Jones, Pat Kociolek, Mark Hunter, and others have included me in dinner conversations about ecology and ecological interactions.

From one perspective, I am a scientific mutt. I was trained as a marine biologist working in biomechanics and animal behavior, have done addition training on the neurochemistry of mental disorders and drug addiction, dabbled in neurophysiology of the rat trigeminal system, and now am becoming an ecologist. Each of the groups above has been gracious enough to allow a scientist with a different background to be a part of their conversations. To all of them, I am grateful for their gracious time and expertise. All their thoughts and conversations form the giants whose shoulders I am standing on.

Central to the final product of the words on these pages is the Laboratory for Sensory Ecology. This is a group of individuals who have attached themselves to me as graduate students. They aren't giants (yet), but are outstanding sounding boards. They endure my excited pronouncements about a new study and questions about whether this analogy or that thought works. They are brave enough to tell me when the train has left the tracks and smart enough to read this book as both intellectuals in the discipline and as outsiders looking in. They have read every word in this book and have guided my thinking, writing, and editing. Without their input, there is no book. Unfortunately, I have to just list them alphabetically, but each has provided a perspective that is different from every other member. I hope they can see their thoughts in the book and recognize the depth of contribution that each of them has made. I am proud that they are willing to claim me as an advisor. They are David Edwards, Ana Jurcak, Maryam Kamran, Sara Lahman, Tim Ludington, Alex Neal, Kathryn Rapin, Juliet Slutzker, Molly West, and Sarah Wofford. My undergraduate students have also read the book and provide me with sets of eyes that are most closely aligned with my intended audience. They are Elli Kantonen, Florence Montarmani, Zach Morris, and Kaitlyn Trent.

I would like to thank Janet Slobodien and Eric Hardy of Springer for bringing these thoughts to print and for taking a chance on this book. Finally, a constant source of inspiration and joy is my family. Without their help and support, I wouldn't be in a position to write a book. Jen, Meghan, and Connor, you have been essential to my sanity through the scientific journeys of my life.

Contents

List of Figures

Chapter 1
An Introduction

The Hidden Odor World

It is an early spring morning in Northwest Ohio as I prepare for my morning run. The air still has the crispness and moisture that reminds me of the past winter, but there are signs of warmer days ahead. Having rained last night, there is a hint of freshness to the smell of the morning air. These signs of spring will help me pass the time as I follow my normal route. As much as my body longs for the comfort of my bed, I force myself to set off at my usual slow pace.

As I approach the house on the corner, I see brilliant red Darwin tulips are emerging from their winter slumber in my neighbor's garden. Sprinkled in and around the red are bright yellows, deep purples, and oranges, reminiscent of a living Monet painting. Standing tall like a multi-hued army, these colors serve a far greater purpose than to please this runner's eyes. The petals act as showy advertisement to the local insect community. "Come and sample the sweet nectar" they call. In exchange for the advertised meal, the diner will help spread the pollen needed to propagate the next generation of flowers.

Looking around, I can see other signs of spring which can be found in the trees. Small green buds are appearing and beginning to add a touch of contrast color to the brown branches. A flash of orange and black crosses in front of me and lands on the oak tree to my left—an oriole sporting its bright vest and black coat looking for nest material. The proud cardinal stands out from the green shrubs wearing its formal dress just like its human namesake. I enjoy these colorful distractions on my journey. They take my mind off of my heavy breathing and stimulate me to think about nature and biology even in this small urban setting.

Colors are not the only stimuli that accompany me on my run. Above the din of my footfalls, nature greets me with her melodies, tweets, and chatter. Half a mile down the road from the cardinal and oriole sightings, I hear the familiar coo of a mourning dove that often greets me on my lonely runs. The mourning doves often sit on the power wires across from my house and softly call as I stretch out my stride to warm my muscles for the rest of my run. As most of the world around me sleeps, we share a connection only early risers can feel. A haunting sound, but I have come to welcome their song as an integral part of my morning routine. A loud and angry

P.A. Moore, *The Hidden Power of Smell*, DOI 10.1007/978-3-319-15651-4_1

argument wakes me from short introspection. Two squirrels chatter to each other as a brown squirrel chases a black one up, down, and around a large Maple tree. As a behaviorist that studies aggression, I am fascinated by their interactions. Their loud clicks are a sure sign that one is an interloper being forcefully escorted from the other's territory. Despite an intense interest to stop and watch, I move on.

At the end of the road, I leave the hard pavement for a small path leading through a stand of trees to a protected meadow. This is the favorite part of my morning ritual. Although I live in a small town, in these woods, I am immersed in the natural world. There are no sounds from cars, cyclists, or people leaving their homes for a morning commute. The only sights or sounds are those produced by nature herself. As I attempt to run more quietly on the dirt and wood chip path, I can focus more closely on the natural world around me. About 300 yards into the forest, I hear a crash followed by a series of slight footsteps. Over to my right, I catch a glimpse of a young deer. Upon hearing my heavier footfalls, the deer freezes. As we catch each other's eyes, we have a long staring contest. I imagine that the deer is attempting to figure out if I am dangerous threat or simply something to ignore. The outcome of that decision is clear as the graceful deer dashes away from me, easily clearing any downed trees in its pathway.

I emerge from the forest to the bright sunlight that has engulfed the small meadow. The ground is still muddy from a rain 2 days past, and there are patches of standing water forming small aquatic ecosystems. Although a petite meadow by any measure, the meadow is teaming with life, and I immediately notice one of its most familiar members. Perched high atop the large reeds that serve as sounding posts is a red-winged blackbird. Named for the brilliant sergeant-at-arms patches on its shoulders, this beautiful bird is performing a common spring ritual for many animals; the male is trying to establish a home territory in hopes of attracting a female and raising a family. As I continue along the path, the bird slowly hunches its back, displaying even more prominently its red stripe, and gives a loud "braaaaack" call. Off in the distance and out of my line of sight comes a return response. Another blackbird has heard the first call and is attempting to negotiate where to draw the line on their adjacent territories. The bird closest to me quickly flutters over to another reed. Presumably closer to his competitor. Over and over again, these two engage in a series of calls in an attempt to have as large of a territory as possible. With each call, the birds are hoping to warn other males that this area is its home and they should stay away. Other birds have begun to respond in kind to the first two and soon the whole meadow is alive with their songs.

As my legs begin to tire and my breath becomes ragged, I sadly leave this little paradise and begin my trek along a series of sidewalks that lead to my own home. The return trip is far less stimulating, providing me with ample opportunity to reflect on the morning's sights and sounds. The springtime sounds of birds singing to establish territories or attract mates, the slow greening of the world as the trees awaken from their winter slumber, and the colorful show of flowers are the signs that indicate spring is upon us.

As visual and auditory beings, we humans relish these springtime harbingers. The bright red tulips and the calls of the blackbird remind us that warmer days are

Fig. 1.1 Running meadow of odors

coming. These visual and auditory cues signal time to pack away the winter coats and find lighter clothing to enjoy the warmth outside. Similar to the tulips and blackbirds, we communicate with each other using words, phrases, and sounds. Music and language are key elements of human society. Images on the television, in movies, and in photographs are used to convey information, evoke feelings, and entertain. For most of us, we perceive the world through our eyes and ears. Our other senses, such as smell, taste, and touch, act as supporting cast members to the leading roles of hearing and vision. The natural world, such as the one experienced on my run, often brings these minor players to the forefront. Yet, to treat our noses and tongues as second class sensory citizens is like walking through The Louvre with sunglasses on. Imagine studying the Mona Lisa through polarized shades. Certainly, one could see the painting, but the total sensory experience is undercut unless all of our vision is used.

Just imagine that we have invented special glasses that give us the power to see the odorous world the way that other organisms perceive it. Put your pair on and walk outside for just a moment. As the bright sunlight hits our eyes, we would encounter a world far different from what we would normally expect. The air is full of molecules carried by breezes. Chemical signals would flood our eyes just as surely as sounds overwhelm our ears at a cocktail party. Stare at any plant and you would see compounds being released into the air from leaves, bark, and roots. A squirrel in a tree exudes carbon dioxide and other compounds with each breath. Glance along its brown body and notice that specific points (scent glands) appear to be slowly releasing chemical signals. If we could translate these signals into language, we would see phrases, sentences, statements, songs, and other messages waiting to be intercepted and interpreted. Taking in the scene as a whole, we would see a symphony of chemicals being played by the creatures of nature in their daily conversations (Fig. 1.1).

One of my favorite examples of the symphony (called chemical communication) comes from the work of Dr. Michael Breed and his coworkers at the University of

Colorado. For years, he has studied various forms of recognition in social insects. (The reasons why these insects need social recognition are explained in Chap. 6.) Focusing on Neotropical ants in Panama, Dr. Breed has discovered that these ants have sentries posted at entranceways into their colonies. The sentries perform a thorough "smell" check of each ant that enters the colony. In many ways, these ants have evolved their own TSA agents whose smell check is as thorough as any whole body scan at any airport. If the ant passes the test, it is allowed access to the colony, but if the wrong odor is detected, the sentries forcibly carry the offending individual out of the anthill to the outskirts of the colony's territory. Each colony has its own odor, called a "label," and the odor can arise from a genetic basis; it may be given to the workers from the queen ant or may arise from specific foods that are present in a particular colony.

The genetics of social insects are rather interesting because only the queen produces eggs. The entire colony is related genetically. The unique nature of the ant's genetics doesn't stop there. The queen of the colony produces two types of eggs. One set of eggs are fertilized and produce females and those not fertilized produce males. All males within the colony will have an identical genetic makeup and all of the females will share at least 50 % of their genes. Different colonies will have different queens and, if the recognition pheromone is based on the genetics of the queen, then the colony will have a unique odor signature.

This nestmate recognition serves to keep intruders out and helps to ensure the survival of the colony. With patience and keen observation, Dr. Breed and his coworkers noticed some ants taking food out of one colony and transporting it to a nearby colony. These ants were discovered to be thieves that were stealing food from one colony and taking it to their home colony. But how could these thieves penetrate the chemical recognition of nestmates performed so thoroughly by the sentries? It appears that through repeated attempts to gain entry and the subsequent rebuffs by sentries, the nestmate chemical gets transferred to the interloping ants. When enough interactions with the sentries have been performed, the thieving ant smells sufficiently like a nestmate ant to the guards. This acquired chemical camouflage now allows the intruder ants to sneak in, take food, and secure safe passage back to their home colony.

This chemical camouflage and name tags are just a glimpse into the hidden world of chemical communication so prevalent in nature, and yet we are oblivious to it. Chemical communication is an ancient art, first performed when two single celled organisms exchanged chemicals in the primordial soup. Through the ages, the use of chemicals as a form of communication has proliferated to such a degree that there is not a single organism on earth that does not use some aspect of it. This can be attributed to the chemical nature of life itself; every single organism must take in, transform, produce, and release chemicals. Consequently, nature is full of chemicals, all with secret information about the inner workings of the organism that produced them. What species is it? What sex is it? Is it healthy, sick, dominant, or ready for reproduction? These are just a few of the questions that can be answered by smelling the world. Every organism is capable of detecting and responding to chemical signals, and they only differ in the degree to which chemical communication plays a role in gathering and transmitting information.

It is impossible to make such lofty claims about the nature of any other sense. There are numerous examples of organisms that do not use vision, from cave fish to bats. The ability to produce and receive sounds occurs in a surprisingly few number of organisms. Virtually all of life uses chemical signals. Even the lowly bacteria communicate to each other chemically. But despite the importance and prevalence of chemical communication through nature, the richness to which organisms have developed the ability to communicate using signature molecules is often underappreciated, even within the academic community. A recent informal survey of the latest textbooks in animal behavior, physiology, neurobiology, and ecology shows that there are 20 times as many pages devoted to vision and hearing as there are to chemical communication.

To illustrate how extensive and diverse nature's chemicals are, let us return to my morning run, but this time with our special glasses on. Retracing the same steps, our glasses should reveal to us our chemical world. Remember those red, yellow, and purple tulips, brilliantly calling to the insect world? Those colors represent only part of the picture. As we know, many flowers produce exquisite scents. And in spite of what we might like to think, flowers did not develop those scents to add ambience to our homes, but rather as a way to communicate to the insect world that great bounties can be reaped within their petals. Flowers have developed scents to attract all types of creatures, such as bats, flies, bees, and moths. As we shall see in later chapters, an odor that is irresistible to a fly or bat would not make a good scented candle used in our homes. The Titan Arum flower, for example, has evolved a scent similar to that of rotting flesh mixed with burnt sugar. A distinctive odor that is hard to imagine as a fragrance. The flower uses this foul odor to deceive sweat flies that normally lay eggs in carrion. The Titan Arum's "perfume" serves to attract the flies to its massive bloom in order to ensure pollination and continuation of the species.

My jog continued past the gardens and on to the small forest before the meadow. This small forest is a mixture of trees dominated by oaks, often used as symbols of strength and might. Yet the massive plants can be devastated and denuded by tiny insects. Unable to defend themselves, how do our mighty oaks fend off these voracious attacks? Once an oak tree has been attacked, the tree begins to produce a distasteful chemical sequestered in its leaves. This compound, often called a secondary metabolite, makes the leaves less desirable to hungry insects driving them to other trees. Although if you are an oak tree within a large forest, there is a good chance that the neighboring trees may be related as either your offspring or your parents. What good does it do to send off all of your insect enemies to eat your family and friends? Obviously, the oak tree should tell the neighboring oaks about this potential threat, but how do you communicate if you are a tree? Nature again has solved this dilemma using chemical communication. The oak tree that has been attacked, not only produces a compound that makes its leaves less palatable, it also produces a compound that is sent through its roots to the surrounding trees. The surrounding trees detect the presence of this compound and begin to produce the anti-herbivory chemical in their leaves also.

Remember that young deer running away from my footfalls? A quick glance with our new glasses reveals a whole array of chemical signals. Carbon dioxide, a sure sign of life, emerges from its nostrils. Just below the eyes, the forehead glands emit chemicals. On its feet are interdigital glands, and on the hind legs the metatarsal and tarsal glands. Finally, as the deer bolts away from us, we glimpse anal glands. If we include urine as a chemical signal, a quick tally reveals the deer's six different sources of chemical cues. These signals function for individual and offspring recognition and sexual receptivity. As we leave our chemically conversant trees and deer, we approach the meadow and the calling blackbirds.

If we look closely among the blades of grass, reeds, and other plants that line the meadow path, we will find numerous ticks. Far from living the glamorous life, ticks must suck the bodily fluids of other organisms, mainly mammals and birds, to sustain themselves. It is a challenging lifestyle. How do you locate a large, quickly moving, suitable host when you yourself are of miniscule proportions and relatively slow moving? Imagine being a tick and walking through a redwood-sized forest of grass blades searching for a deer, mouse, or other potential meal. If you are lucky enough to chance upon a worthy candidate, how are you going to position yourself to bite into the soft, fleshy portions of your new host? If it is a deer quickly passing by, all you have in front of you are impenetrable hooves. Ticks have solved this problem using patience, ingenuity, and chemical signals. They will slowly climb up the nearest tall reed, blade of grass, or branch and will hang themselves upside down. Sitting in this position for potentially long hours to days, they will wait until they catch a whiff of an approaching organism. Once the right chemical is sensed, they release themselves from their perch and drop on the unsuspecting host. At first thought, you might imagine that the chemical that releases this behavior is some complicated and important organic molecule signaling a healthy meal, but you would be mistaken. The signal is nothing more than carbon dioxide, CO_2, a chemical that most organisms emit constantly. It is a simple, and yet powerful compound that tells the tick, "A warm-blooded host is near. Let go and drop to attack."

These are just a few of nature's chemical communication techniques. Humans have "chosen," through evolutionary adaptation, to perceive our world mainly through sights and sounds, but these channels do not often offer the best viewing options. For example, imagine that you are a crayfish sitting on the bottom of the mighty and muddy Mississippi. Light is a very limited resource and provides little in the way of a description of your environment. In certain sections of the river, torrents of water are rushing by you creating a wealth of background noise, literally drowning out any possible auditory signals. How are you going to find food, a mate, or a safe haven? These important problems are often solved using chemical signals. Before we can answer these questions and understand the role of chemical signals in nature, it is important to understand what capabilities organisms have with respect to chemical signals.

1.1 Can't See Past My Own Nose

Before we delve any further into the world of odors, some definitions and explanations are in order. Olfaction is the more formal name for smell and although this book is titled "The Hidden Power of Smell: How odors control and influence our lives," olfaction is just one of the many channels through which organisms engage their

Fig. 1.2 Noses

chemical world. As humans, we focus on two distinct senses: olfaction (smell) and gustation (taste). In reality, these two senses are not as distinct as thought at first glance.

When discussing an animal's ability to detect chemicals, up to four different senses can be included. Olfaction is the most prominent and certainly the most predominant in our mind. When we think of smelling, we usually envision odors being pulled into a bifurcated nose by a sniff. These are chemicals that are brought to our nose by wind or air movement. Often thought of as the distant sense because the odors are originating from a spot far from our nose. Although this scenario is true for most mammals, it is quite mammocentric meaning that this concept or definition doesn't fit most life forms. (Mammals are an exceedingly small group of life on this planet despite our focus and fascination with this group of animals). When defining mammal's olfactory abilities, words such as airborne, sniffing, and nose really don't fit with how aquatic animal or insects smell their world. Arthropods (armored animals) including insects, crabs, lobsters, and spiders use bifurcated antennae and move rapidly through the air or water to detect chemicals near or far. Crustaceans, as a subgroup of arthropods, have multiple appendages on their head that function as "noses" (Fig. 1.2).

The second sense included in our discussion is also very familiar to us, the sense of taste. Again for mammals, taste (or gustation) occurs when substances enter the mouth and contact the tongue. Small taste buds, which look like the letter "D" rolled onto the curved side, are all over the tongue. These taste buds are sensitive to a variety of compounds and the familiar and out-of-date conception of four flavors: salt, sweet, bitter, and acidic. There is a fifth additional taste to add to the earlier four and is called umami. Derived from the Japanese, for delicious taste, this taste is often described as "meaty" taste. Summarizing all of these factors together, a concept of mammalian taste has five distinct flavors, is carried by taste buds that are shaped like bowls, is used for close contact (as opposed to distant olfaction), and functions in the consumption of solid and liquid food. Other animals have a very different structure to their taste systems. Fish, for instance, have very similar taste buds, but their entire body is covered with taste buds. Fish also have a taste system located within their mouth similar to humans. Insects like flies don't have taste buds like humans and actually taste with their feet. Crustaceans also taste with their feet (walking legs) and have another taste system on appendages near their mouths. The taste systems for fish, insects, and crustaceans are sensitive to compounds other than the classic five mentioned above.

At this point, we end our familiarity with the chemical senses and move on to more hidden senses that can play a big role in our chemical perception. When we bite into a chili pepper or chew cinnamon gum, we are actually activating a sense that is separate from taste; the trigeminal sense, which is part of our somatosensory system. The somatosensory system includes nerves all along our body that are sensitive to heat, cold, touch, and pain. This is the system that is activated by the menthol in athletic rubs used on sore muscles. For a nice demonstration of the differences between olfaction, taste, and the somatosensory system, find a friend and buy some cherry, raspberry, and cinnamon jellybeans. All these beans are roughly the same red color, so color does not help in differentiating the flavors of these beans. Tell your friend to hold their nose with one hand and then hand them a cherry and raspberry jellybean. If they continue to hold their nose while they chew on each bean, they will be unable to tell the difference between the two; yet they both will taste sweet. This is taste. Now hand them another cherry and raspberry and have them unplug their nose halfway through chewing. The cherry or raspberry flavor will rush out through the back of their mouth and tantalize their nose with cherry or raspberry. This is olfaction. Finally, have them plug their nose, yet again, and now hand them a cinnamon jellybean. Even with their nose plugged, the recognizable cinnamon "taste" will be present. This "taste" is actually a burning sensation that is carried through the mouth's trigeminal nerve.

The final sense is a specialized chemical sense that is currently only known in some vertebrates. It has been found in reptiles, some amphibians, and mammals, but is lacking in fish and birds. This is called the vomeronasal organ, a key sense for many social and reproductive behaviors in mammals. This organ is a small patch of chemically sensitive receptors located above the vomer bone in the roof of the mouth. Animals, such as snakes, goats, and horses, use their tongue and lip movements to place chemicals directly on the vomeronasal organ. Whether humans have a functional vomeronasal organ is an interesting and debatable prospect. This organ is the structure that is receptive to many reproductive pheromones and has been studied extensively in hamsters and other small mammals. Within snakes, the vomeronasal organ is receptive to chemical signals that convey territoriality as well as food.

1.2 Why Chemical Signals?

Many animals have wonderful visual and auditory capabilities. Bats have an amazing ability to locate small rapidly moving objects using only their highly developed hearing. Researchers have placed bats within rooms with transparent fishing line

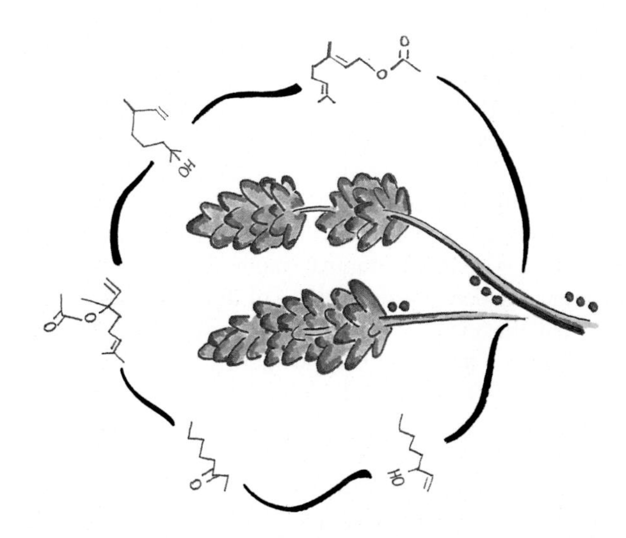

Fig. 1.3 Molecules

strung around the room and the bats have no problem avoiding all the lines. We often marvel at the sonar ability of whales and dolphins and their capability to communicate halfway around the world with their songs. Many birds of prey have outstanding visual perception and can target prey from great distances. With flicker fusion rates well above 100 Hz (this is a measure of how fast still pictures need to pop into and depart from our visual field to give the appearance of a moving image), birds of prey can track small, but fast moving creatures on the forest floor. With our flicker fusion rate hovering around 16 Hz, we would just see a blur of movement if a mouse ran in front of us. Given these examples of the sensory capabilities of animals, we could ask ourselves why chemical signals are used at all.

These signals have specific properties that give them distinct advantages over visual and auditory signals. Longevity, specificity, the ability to alter the physiology of the receiver, and lack of large-scale deceitfulness are among the properties that give chemical signals advantages over other forms of signals. Specificity or a lack of ambiguity of information is probably the key property found in chemical signals and can be formally defined as the uniqueness that a signal provides to the receiver. The visual components of our faces used in facial recognition is an example of specificity. The color, placement, and shape of our eyes along with the features of our nose and mouth provide a fairly distinct visual image of ourselves to the outer world. Contrast this to person's hand. There is little about a hand that signals a singular person. Yet, the rarely is there any confusion when studying the features of someone's face. There are millions and millions of different types of chemical compounds in the world. Each one of these is a potential signal and for the most part, each one has the potential to activate a receptor cell. Most chemical signals are organic compounds with a multitude of shapes and sizes (Fig. 1.3). Many of these compounds have side chains that add specific chemical properties that fine-tune their role as signal compounds (explained in Chap. 2). Alter the chemical structure just slightly and now the chemical takes on a whole new meaning in nature.

Male moths take adventurous flights to locate females willing and ready to mate. Often in a given area, female moths of many different species are all releasing their alluring perfumes in unison. How does the male moth make the right choice and follow the perfume from his female? Specificity of the signal is the answer. A subtle change in the shape or size of the pheromone alters the meaning of the message dramatically. This is similar to comparing the words "pheromone" and "Pheromone." These words have the same meaning to us. Yet, a small difference in two pheromones is all that is needed by our male moth. The ability for molecules and their shapes to be specifically tied to a single piece of information is unique among the sensory world. We find that many chemicals have one specific unambiguous meaning.

Compare this specificity with other types of signals. One of the more common visual signals found in nature is red coloration. Nature is full of various red hues in flowers, on birds, insects, fish, and mammals. The red underbelly of a stickleback fish signals readiness to mate. The red on the wings of a red-winged blackbird serves as a signal to other males. The red on a baboon is used to convey internal moods such as receptivity or anger. In our own world, red can mean stop as in a stop light, caution if used on a warning sign, a good wine of the port variety, and pain if we see red skin around a wound. Colors and shapes provide less specificity than chemical signals.

Words also often have multiple meanings. Listen to a politician's speech with the subsequent analysis from the other political parties, and it is all too clear that words have multiple meanings. Subtle changes in the words used in a phrase and even tone used while speaking can alter the meaning of certain words. We can tell where someone is from based on the accents used during speaking. The British pronounce the words "vitamin" and "schedule" different from Americans. The different dialects and words used in Spanish vary across the many different Spanish speaking communities. In a similar fashion, humpback whales have different accents or dialects that allow other whales to distinguish whom is speaking. Over time, some of these errors or differences accumulate within local populations producing regional differences. Within both acoustical and visual signals, these regional differences in signal meaning add a level of ambiguity that is absent from chemical signals.

In addition to specificity, other properties are unique to chemical signals. Longevity refers to the length of time that the signal remains in the environment. Many mammals use chemical signals to mark territories; for example, canines use urine as a sign of territory boundaries. Marking of territory by the use of urine signals is the precursor behavior to our domestic canines' need to urinate on fire hydrants. By using urine, as opposed to howls or visual displays, the canines, such as wolves, ensure that the territorial signal remains even after they have left an area. Sometimes these signals are detectable by other animals in the area for days and weeks later. Some animals, like the sea lions and red-winged blackbirds, use fierce vocal and visual displays to mark and establish territories. These animals often expend tremendous amounts of energy in continual displays at all ends of their territory. Wolves mark and move on without worrying about intruders into their territories. If a new wolf comes into the territory, they will smell the urine marks and will instantly know that another wolf is around.

Most visual and auditory signals get processed by the conscious part of our brain. This means there is at least a little thinking moment before we respond to those stimuli. Words in the form of poetry, music, and art move us emotionally. The crescendo of Gounod's Ava Maria has the ability to send shivers down the listener's spine, and the first viewing of Michelangelo's Sistine Chapel leaves one awestruck and full of wonder. The vivid sights and sounds of our world have the ability to touch us in such a way as to alter our moods. Although they appear spontaneous, these emotional and touching responses are, in reality, a highly processed response from our brain. These stimuli are perceived by our eyes and ears, which subsequently pass the information on to the conscious part of our brain. From here, our brain searches for the appropriate response from the suite of learned behaviors. Conversely, chemical signals can bypass this conscious perception and act directly on an organism's emotional and physiological state. This process is so powerful that it can alter fundamental physiological processes.

One of the most severe examples of chemical impacting the physiology of another organisms is seen in the chemical castration and abortion found in social mammals. Mice are very social animals, and if several female mice are housed together their estrous cycles will become very irregular. Now if a male mouse, or just its urine, is placed in the same housing, the estrous cycles will eventually become synchronized (called the "Whitten Effect"). Within mice, if one of the females is impregnated, additional chemical signals can influence the development of the embryo. If the paternal mouse is removed and a new male is placed with the female or even if urine from a different male is introduced into the housing, chemical cues from these new sources can cause the pregnant female to abort the pregnancy. From an evolutionary point of view, aborting a pregnancy from a previous male is beneficial to both the new male and formerly pregnant female. The new male has a chance for reproduction now that the female is no longer pregnant and the female has a chance to sire offspring with a superior male. This phenomenon is called the "Bruce Effect" and is a powerful example of the control that chemical signals can play over the physiology of another organism. Not just limited to mammals, it is possible to find odors that control the reproductive status of bees or even the development of bees into queens or workers. As these examples show, chemical signals can work from outside the body to impact the inside, but the opposite can also occur; chemical signals from inside the body can impact the behavior of other organisms. Examples of this inside out effect will be covered in later chapters.

The final advantage that chemical signals have over visual and auditory cues is that chemical signals can provide a direct link on the internal physiology of the sender. Female mammals produce different odors depending on whether they are in estrous or not. Female moths produce mating pheromones when they are ready to mate. Crayfish produce different body odors whether they are a dominant or subordinate crayfish. Even changing dietary consumption has the potential to change the types of chemicals produced by an organism. It is often said that you are what you eat. Catfish are wonderfully sensitive creatures when it comes to detecting odors in water. The reason these animals are called catfish is the prominent "whiskers" that protrude from their lower lips. These whiskers are actually called barbells and are covered in taste buds. Almost all catfish use these taste buds to locate food items that may be buried in the sand or silt on the bottom of lakes and rivers. In addition to their barbells, their body is also covered in taste buds. Essentially, we can think of them as giant swimming tongues. Drs. Bruce Bryant and Jelle Atema became interested in whether catfish can recognize each other just using their sense of smell. Catfish are fairly aggressive fish and will often establish a dominant and subordinate relationship if placed in a tank with another catfish. Drs. Bryant and Atema placed two catfish in a tank and watched their behavior. At first, the two fish were quite aggressive toward each other, but soon a relationship developed where one fish was clearly dominant. This fish would swim over to the subordinate animal and would periodically bully that animal. If the subordinate catfish is removed and replaced with a different catfish in a couple of days, the two catfish would begin fighting to reestablish who was the boss of the tank. If the subordinate catfish was removed and replaced in the same tank in a couple of days, the bully catfish would quickly recognize the old catfish and no new fights would occur. This clearly showed some sort of recognition and memory on the part of the bully catfish. But there is a twist to our story. To investigate the role of chemical signals in this social interaction, Drs. Bryant and Atema removed the subordinate catfish and fed the subordinate a different diet over a couple of days. After this new treatment, they placed the original subordinate back into the original tank. The bully catfish reacted to the subordinate as if it were a new catfish. They began to fight and to reestablish their dominant relationship. Simply by changing the diet of the animal, they were able to change how one catfish perceived the other. The chemical signals produced by organisms originate from the basic building blocks of chemicals within their bodies. The ultimate source of these chemicals is the foods and other matter that is consumed on a daily basis. If you change the diet, you change the availability of those basic building blocks.

1.3 Why a "Hidden Sense"?

As stated above, we are predominantly visual and auditory creatures. We use these sensory channels to communicate to each other and to send messages across many different media. Our TVs, smart phones, internet, driving signs, and radios are strictly aimed to maximally stimulate our visual and auditory senses. When purchasing these electronics, we are bombarded with a host of terminology that tells us the pixel resolution or the auditory range of the speakers. The best movie experiences are high definition 3D affairs that ignore our sense of smell. Movie screens are getting larger with higher resolution while the speaker systems emulate a surround sound auditory environment. Even our everyday nomenclature is fraught with references to these two senses. "Do you see what I mean?" or "I hear you" are used to connotate understanding. In psychology and biology, we run blind or double blind experiments which has nothing to do with the functioning of our eyes. This phrase refers to the concept that the researcher performing part of the experiment is unaware of the experimental setup in order to minimize bias. Our culture and society is dominated by the use and referral to these two senses.

As explained in Chap. 2, we don't usually "think" about the chemical senses in the same way that we think about other signals. During my summer field research season, I often take time after dinner to sit outside my cabin and do some reading. Reading *Walden Woods* while sitting near the shores of Douglas Lake surrounded by 10,000 acres of woods allows me the time to think about Thoreau's words and thoughts which are transmitted by the words on a page as seen by my eyes. If I am writing or editing papers, I often put on some music to stimulate the creative thought process. Listening to Joshua Bell on *The Red Violin* soundtrack often does the trick for me. Although occasionally I do stop writing for periodic moments to close my eyes and really focus on the music. These signals, the words on the page and the sounds from my speakers, and the information they convey are processed differently in our brains than chemical information (see Chap. 2). Consequently, one doesn't light a scented candle to think about the chemicals emanating from the wax and wick; we use these scents to set moods or touch on emotions. When we walk by a Thai restaurant and smell the sweet aroma of coconut milk mixed with the spiciness of curry, we don't pause to consider the structure of coconut trees with the color of curry leaves. The aromas of our world remain "hidden" from the conscious part of mind and often play a background role in our lives.

In surveys about which sense would a person rather lose, your sight, your hearing, or your sense of smell, our chemical senses are almost universally chosen as the first sense to lose. This is only if the creators of the survey even include the sense of smell. Even when paired with the sense of touch as in "would you rather lose your sense of touch or smell," olfaction comes in a close second. These informal surveys reflect a common thinking that olfaction is a secondary sense to living our daily lives in a modern world. Yet, as much as chemical signals are really a background sense for most of our daily lives, a simple trip into any bathroom might tell a different story as a glance around the room would reveal scented soaps, shampoos, perfume, cologne, flavored toothpaste, mouthwash, and air freshener. Walking into a kitchen would reveal shelves full of spices such as salt, pepper, cinnamon, thyme, or garlic. Without our sense of smell and taste, a potato and apple are perceived as identical to the consumer. I, myself, would rather not think about a snack of smoked Gouda with a dram of Oban or Talisker without my taste buds or nose working properly.

1.4 A Language of Odors

The word "language" can have a variety of meanings that depends largely on the background and profession of the person proposing the definition. A casual glance at several different dictionaries shows that a language is defined as the communication of ideas and emotions using sounds or words or a systematic method of communicating feelings, ideas, thoughts, or emotions through speech, either written or spoken. This is probably what most of us think about when we talk about a language. It is our words and how we use our words that convey our information to the person to whom we are speaking. This definition has a problem in that it excludes nonverbal communication, such as sign language. After searching through seven dictionaries, I find a definition that quickly becomes my favorite: Language is the transmission of emotions or ideas between living creatures ***by any means*** (my emphasis). Any bias toward auditory cues is absent from this last dictionary entry, which begins to open the door for a language of smell and is the reason that this is my favorite definition. All languages have a common set of features. These features can be summarized and used as a guide from which we can judge, in a rather casual way, whether something is a language or not.

The symbols or words of a language are the basic elements that are used to construct complex ideas and thoughts. These symbols or words often have multiple meanings, which are only unambiguous when used with other words in a specific context. Thousands of words or symbols can be made from a much smaller set of building blocks. For the English language, there are 26 different letters. It is the variety of ways that these small building blocks are used that give depth and meaning to our ability to communicate feelings, ideas, or emotions. The same 26 letters that move us so well in a novel such as "The Color Purple" also inform small children in Dr. Seuss' "Green Eggs and Ham." The communication of ideas and emotions are not due to the individual building blocks or letters but in the way that they are used to construct words. In addition, the many words impart different information depending on what order they appear within the sentence and the context in which the word appears. Consider these three letters; a, e, and t. They can be combined to make the words eat, ate, or tea. Although the letters are the same, the meaning of each of these words is different. Even when the word is the same, the context in which the word appears alters the meaning. The word "bat" has three different meanings in each of the following sentences. See the bat eating insects. The player just broke his bat. I used my hand to bat the fly away.

Finally, these building blocks are used according to a set of rules (even if those rules are rather confusing and contradictory). The common knowledge and use of these rules make a language accessible to those trying to communicate. The rules can be as simple as every word needs a vowel or as complex as the construction of topic sentences and paragraphs. Without common knowledge of these rules, ideas cannot be communicated effectively. If I constructed a weird rule, say that every sentence needed to be an alliteration, and you used a rule that every third letter needs to be accented, we would struggle mightily in understanding each other.

If we apply this same analysis to smell, we can find that the features found in our spoken language also appear in a language of smell. Our language is constructed from words, and these words are subsequently constructed of a few different letters. In a similar vein, the many "words" of the language of smell, the molecules of life, are constructed from a very small set of atoms. The immense diversity of chemical signals that will be illustrated in this book are based on a handful of atoms: carbon, oxygen, hydrogen, nitrogen, and to a smaller extent, sulfur. This is the alphabet of the language of smell. Just as a, e, and t can be combined to form many different words, carbon, nitrogen, and hydrogen can be combined to form many different molecules that are used in chemical communication (see Chap. 2).

In our language, the meaning and context of the words becomes clear only after it is used in conjunction with other words. Within the language of smell, many different organisms use identical "words," i.e., molecules, which can make the transmission of ideas confusing at best. Recent work has shown that most instances of chemical communication rely not on a single molecule, but on a complex mixture of molecules. Indeed, it is the mixture of molecules that is necessary to effectively communicate ideas and elicit the appropriate behavioral response in the receiver. This shows that it is not necessarily the molecule that matters, but the mixture or context in which the molecule is used that conveys a certain idea.

Finally, the most difficult aspect of the idea of a language of smell is the demonstration that this language has a set of rules that are common knowledge to the participants. Because this is such a difficult concept to present without a set of stories that illustrate how chemical signals are used in the realm of nature, I will ask your leave to return to this idea in the last chapter.

1.5 Taking Time to Smell the Roses

The scale of nature is often difficult for our human minds to understand. In his excellent book, *The Black Swan*, Nassim Taleb constructs an argument that our minds have evolved to understand linear relationships, but not logarithmic ones. For example, if we take all of the adult people of the world and line them up according to height or even weight, the difference from the smallest person to the largest person is a rather simple linear relationship. At one point, the smallest man and largest man met and their difference in height was less than 6 feet or about the average height of a human male. This difference is only three times that of the smallest man. This is a nice linear relationship meaning that the differences are within a single order of magnitude (0–10). If the relationships fell beyond a single order of magnitude (0–1000), this is where our mind begins to falter.

If we compare this linear human height difference to biological lifespans, we can begin to understand the difference in logarithmic scales and linear scales. A common late spring and early summer yearly event is the large scale emergence of mayflies. Fortunately, mayflies do not bite and as adult live a mere 24 hours. A short life span compared to our own 80–100 years. Yet, to some of the longest lived creatures on this planet, our lifespan may seem like a flash in the pan lifespan of the mayfly. Some aspen colonies that are estimated to be as old as 80,000 years and some species of fungus are estimated to be over 10,000 years old. Comparing the mayfly lifespan to that of humans, there is roughly a 37,000-fold increase in the human lifespan, so this is a 4-fold order of magnitude difference in length of life (order of magnitude refers to the number of power of tens is necessary to equate the two measures). A larger order of magnitude difference appears when comparing the mayfly lifespan to that of the aspen colony. Here the difference is over 29 million mayfly lives or a sevenfold order of magnitude difference. Thinking about a 3 times difference (such as human height) is far easier to comprehend than thinking about a 29 million-fold difference. The geological time scale or size scales in the universe are also logarithmic in nature. Sensory systems, in general, and chemoreception, in particular, operate on a logarithmic scale as the intensity of stimuli in nature vary greatly.

Because logarithmic differences are so hard to imagine or think about, sometimes we are impressed by numbers that really aren't that impressive. Sharks are highly sensitive to blood and use the scent of blood to locate and hunt prey. Their sense of smell can detect approximately one drop of blood in a million drops of ocean water. At first glance, this sounds impressive and is about halfway between the 3 times height difference in humans and the 29 million-fold difference in live spans. This ability is even more impressive when we think about movies like Jaws or about the killing power of the predatory Great White Shark. In reality, sharks' olfactory abilities are quite pedestrian compared to other creatures. Many aquatic crustaceans can sense concentrations one million times smaller than sharks can. So one drop of food in one trillion drops of ocean water. The exact thresholds for canines are unknown, but are probably better than the trillion sensitivity of lobsters and crabs. Some estimates are set at one quintillion. Humans likely fall around the one part of an odor in a million parts of air roughly comparable with the shark estimates.

Apart from sensitivity of our detection of odors is the range of odors that we can smell. For years, scientists and perfumers alike thought that humans were sensitive to about 10,000 different odors. This estimate was in part based on the number of different genes for odor receptors that had been found. Interestingly, this number became a sort of dogma and was never really tested. That is until quite recently when scientists at Rockefeller University teamed up with scientists at Howard Hughes Medical Institute to really test this assumption. The lead scientists, Dr. Leslie Voshall and Dr. Andreas Keller, asked human participants to play an odorous equivalent to the three card Monte game. Dr. Voshall and her team would provide participants with three vials of odors. Two of the vials contained identical odors where the third vial contained a different odor. In the three card Monte game, marks are asked to locate the queen amongst three cards. Now, Dr. Voshall was not attempting to trick participants out of money, she only wanted to ensure that the task was difficult enough to determine how many different odor combinations humans can actually smell. The interesting twist on the usual "sniff a vial and identify the odor" human olfactory test was that these researchers combined up to 30 different single odor molecules in the vials. So rather than ask participants if they can identify an odor, they asked the participants to identify the vial that was different. From this study, Drs. Keller and Voshall have provided a conservative estimate that humans can distinguish about one trillion different odors; a significant increase over the previously thought of 10,000 odors.

Compared to the range of colors (7.5 million upper estimate) and tones (approximately 340,000) that we can distinguish, our olfactory system shines the brightest among our senses. When asked about the amazing range of human chemical sensitivity, Dr. Keller responded, in an interview with Time Magazine, "We just don't pay attention to it and we don't use it in everyday life." So, when we are admonished to slow down and smell the roses on our daily journeys, we aren't just taking time to enjoy our life, we are really bringing to the forefront of our mind the most powerful sense we have. By becoming aware of our abilities within the olfactory realm, we are revealing the hidden talent that lies within our nose.

1.6 Our Cousin's Perfume

One of our most closely related species on this planet (and unfortunately endangered) is the lowland gorilla. Depending how one defines relatedness, current estimates on the differences in the human and gorilla genome range between 1.75 and 1.37 % difference. A small difference given the number of genes in our shared genomes. Gorillas, like humans, are social animals and live within groups of varying sizes. There is sexual dimorphism in this species with males being significantly larger. Gorillas live in groups that are fairly stable over time and are led by an adult male for a significant period of time sometimes for more than a decade. The older males will accumulate grey hair on their backs as they age and hence are termed silver-backs. These males exert social control over the group and determine when the group eats, goes to sleep, and moves within their home range. In addition, the males mediate dispute and have sole access to the reproductive aged females within the troop. A lot of work has shown that gorillas have a fairly complex language and that this language is used to communicate within the troop. The famous field biologist, Dr. Dian Fossey, played a significant role in elucidating the vocalizations among the gorillas troops.

Most likely because of the closeness between our two species, our reliance on sights and sounds for communication, and the lack of a functional vomeronasal organ, there is a dearth of information on the role of chemical signals in the social communication of gorillas. Of all of the primates, humans, chimpanzees (another close relative), and gorillas have the highest density of secretory glands in the arm-pits. So, gorillas are certainly equipped to produce body odor in a similar fashion as humans and are more equipped to do this than most other primates. One of the first rules for attempting to understand the role of chemical signals in any species is locating where potential signals originate. In humans, the underarm is one of those key sources for bodily secretions that could produce characteristic scents. This, of course, leads to the use of underarm deodorants or antiperspirants that would block, hide, or inhibit the nasty odors that would make carpooling, elevator rides, and intimate nights far less enjoyable. In actuality, body odors aren't so much our secretions, but the secretions of the bacterial culture in those armpits feeding off of our signature body chemicals.

The interaction between the bacteria and our secretions is important because as our emotional situation changes, our secretions and body temperature also change. Under high stress or pressure situations, like giving a talk in front of strangers, body tempera-ture rises and sweat production increases. In similar conditions like fearful events, the similar physiological changes happen. All of these changes create the opportunity that our body odor may reflect our mood in any given situation. The cliché "the smell of fear" has at least a kernel of truth to it. Recent work in humans has shown that the odor produced by a scared person can also trigger similar emotional and physiological responses in those that smell that fear (More on this in Chap. 10).

Given this background work, researchers Drs. Michelle Klailova and Phyllis Lee wondered if there was a hidden communication using chemical signals within gorillas. Drs. Klailova and Lee recorded their perception of different odors being produced by the silverback gorilla, auditory calls, and the behavioral responses of this troop. The silverback, named Makumba, ruled over 13 other individuals. Recording the odor production of a large silverback lowland gorilla in the middle of an African preserve is difficult to perform with high precision equipment, so the researchers focused on those times that Makumba produced intense odor signals. With over 3000 behavioral measures and 1000 odor qualifications, the researchers have a good start on a very interesting data set.

So what did they find? Although a very preliminary study, it appears as if Makumba can modulate to some degree the intensity of the production of odors and that the production of odors (at least the intensity of the production) may be context dependent. The most intense odor production occurred when one of Makumba's youngest offspring and mother were not within his presence. Conversely, during more quiet periods for the troop or when there was close contact, odor production was far less intense. At the very least, odor production by Makumba appears to be context dependent, "loud," or intense odor production during stressful situations and "quiet" or low odor production during periods of relatively low stress within the troop. Now, it is entirely possible that the odor production is a physiological result of the stress rather than an active production of a signal. This would be similar to yelling "ouch" as an automatic response to a painful event, such as stubbing a toe. Even if the yell is not aimed specifically as a communication for others to avoid the source of the pain or to come help, the "ouch," if screamed loud enough, does the trick anyway. So, if Makumba produces intense odor signals in response to the stress of an unseen offspring, it is possible that these odors still communicate stress and warning to other members of the troop. This could be the rudimentary basis of a language of smell. The authors conclude that "gorilla adult males appear to use highly context specific chemo-signals to moderate social behaviors" which is what gorilla vocalizations also do.

Previously in this chapter, an argument was laid out for the rules of a language, and context specific communication is one of those rules. As a reminder, the context specific nature of words and letters allows the reader to differentiate the meaning between "bat" the device to hit a ball and "bat" a flying mammal. If the novel work on chemical communication in gorillas (our close relative) holds up under further scrutiny, then a similar line of work in humans could reveal a hidden level of communication that may be more powerful that we currently think. We often think of chemical signals as our unintended and uncontrolled response to a situation rather than an active form of communication. An excellent example of these differences is demonstrated admirably in Patrick Süskind's novel "Perfume" (Das Parfum in its original German version). In this fictional book, the protagonist (Grenouille) is born with no body odor, but has a perfect sense of smell. Think of a mute composer with a perfect pitch for hearing. After becoming an apprentice perfumer, Grenouille uses his perfect nose and knowledge of scents to communicate specific messages and emotions to people around him with odors. Applying these fanciful concepts to real life, a deeper understanding of chemical signals as *both* an emotional releaser and as communicative carriers of critical information would be a significant advance in our current perception of chemical signals.

1.7 The Exquisite Sensation of Sushi

These are just some of the fascinating stories of chemical communication found in nature. Rather than just having a book about human olfaction and taste, by turning to the broader world outside of human society, I can show all of the amazing tasks that nature performs using the sense of smell (and taste, trigeminal, and vomeronasal). Animals and plants use chemical signals as an essential element of their daily activities and are the experts in demonstrating what we, as humans, are missing by not paying attention to the sense that is literally right in front of our eyes. After exploring this hidden world of odors in the aquatic realm for over three decades, I feel compelled to share the stories that nature has revealed to me and many others.

This book is written with the purpose of elucidating the many amazing roles chemical signals play in the daily functioning of nature and to reveal how these chemicals also influence and control our daily lives even if we are unaware of this control and influence. As such, this book is organized around the daily needs of organisms and how those daily needs translate into our everyday lives. These needs include fundamental aspects of our lives such as our need to eat (foraging), the need to be a part of the human society (social interactions), the need to find and build a home (shelter), and our drive to have children (reproduction). I have a twofold purpose in writing this book. First, I want the reader to walk away, at the end of the book, with a deeper appreciation for the prevalence of and complexity of chemical signals in nature. If this increased awareness happens, strolls through a garden, walks down a forest path, a kayaking trip down a stream, or simply eating at your favorite restaurant should be transformed into a different journey. One where the smell of the forest floor after a cool spring rain brings as much joy as the sight of the morning dew on flower petal. Or that stepping into a restaurant is as moving as listening to an outdoor live performance of "Moonlight Sonata" on a summer's eve.

One of my favorite gastronomic events is the consumption of good sushi. Since I have worked in the chemical senses my entire scientific life, I make my sushi eating experience a total indulgence of all of my chemical senses. The first step in my process involves the proper mixture of soy sauce and wasabi. Although this is usually frowned upon by high end sushi chefs, I enjoy adding just a touch of extra spiciness to the sushi. A dab of wasabi is mixed into a small bowl of soy sauce. In preparing this concoction, I have a particular tannish color that I am aiming for. Titrating the mixture of soy and wasabi, I reach the appropriate color and am ready for a fabulous treat. Dipping the sushi (rice side up) in the mixture, I raise the sushi to my mouth, but hesitate for a fraction of a second. At this point in time, I add a little sensory trick to augment my perception of the experience. I close my eyes and mentally shut out all of the sound around me. I force myself to focus solely on the information coming into my mind through my chemical senses. As the piece of Maguro sushi anoints my tongue, I am aware of an overload of sensory information rushing to my mind. The salt of the soy stimulates my taste buds which enhances the other channels of information. The wasabi (most likely green-dyed horseradish) activates my trigeminal sense and a shiver of pain envelopes my body. As the odors gently waft from the back of my throat to the olfactory cells in my nose, I am engaged in a dancing duet of tuna and soy. Finally, as the volatile chemicals from the wasabi reach my nose, the entire head feels like it has opened up and is alive with the symphony of chemicals. After reading this book, I hope that your odorous life is enhanced as much as my sushi dinners.

My second goal for the reader is to uncover the role that chemical signals play in our human lives. The amount of time and effort we devote to producing the right smell seems enormous given the amount of perfumed products found around our house. Add to that total, those products (kitty litter, room deodorizer) we use to hide other odors seem to indicate that we are attempting to be amatuer chemists in our own home. By drawing parallels between our use of sight and sound in our everyday activities to nature's use of chemical signals for identical activities, I will attempt to make the reader more consciously aware of the chemical world around us. The subtle nature of which these chemicals creep into our mind and influence or even control our behavior is incredible even if we are unaware of that process happening. Thus, when we do play chemists in our homes or even in nature, maybe the reader will be a little more informed about what is happening.

I have finally arrived back at the front steps of my house. Winded and tired, any attempt to sample the odorous world around me would be a futile attempt. My breathing is rapid and through my mouth rather than through my nose. A momentary loss of my olfactory abilities, I still enter my house energized. The world of chemical signals lies ahead of me (and you as the reader). While one journey (my jog) has ended, another awaits in these pages as we will travel through many different habitats to visit a menagerie of chemically communicating animals. Although these animals will be a large part of the stories contained within these chapters, our chemically mediated abilities will always be omnipresent in the words of this book. One eye on nature and one eye on ourselves so to speak (or write). So, as I walk up the steps to open the door to my house, we shall walk through these pages to see if we can open the door to our olfactory world.

Chapter 2
A World of Odors

Signals, Cues, and Information

The morning air has a slight bite to it, but despite this early morning chill, I decide to walk into my biology office. My young companion on this journey has a far more intimate knowledge about this world of odors than I do. I am just an interloper only vaguely aware of all of the smells around me. The traveler accompanying me on this particular trek is a 6-month-old Shiba Inu named Cedric. As any dog owner can tell you, smell is how dogs "see" the world. It is the sense that they use to greet each other, "know" where home territory is located, and when food is available. When I take Cedric to the dog park, the other dogs rush over to get a good sniff to see if they know him from a past encounter. As Cedric and I pass any standing object, be it tree, fire hydrant, or parked car, he wants to stop and take a good sample of the chemical milieu.

His behavior points to one of the unique properties of the chemical senses and their signals; odor molecules have a longevity beyond the mere presence of the organism that produced that signal. A common example of this phenomenon occurs when one enters an elevator after a person with a strong cologne or perfume has ridden the car. Those perfume molecules remain within that enclosed area for a considerably long time. Hotel rooms inhabited by smokers, even for a single night, have a distinct smell to them. Garlic lingers in the kitchen long after the last spaghetti noodle has been consumed.

For Cedric and myself, I can clearly define our two main chemical senses: smell and taste. To be simplistic, smell (olfaction) is done through our nose and taste (gustation) by our tongue. When Cedric sniffs a lamp post, he is using his olfactory abilities, whereas when he licks my hand, taste is being employed. These two processes are carried out by fundamentally different receptor types, they are sensitive to disparate classes of chemicals, serve very different behavioral roles for us, and have different active spaces. An active space is the area where a sense can receive information. In humans, our nose picks up distant sources of molecules: The pizza smell down the road, the smell of fresh cut grass across a lawn, the garlic of a spaghetti dinner. Whereas, our taste system has to be in direct contact with the stimulus: an ice tea with a hint of lemon or the sweetness of a crème brûlée. In many

P.A. Moore, *The Hidden Power of Smell*, DOI 10.1007/978-3-319-15651-4_2

nonmammalian systems, this separation is quite murky. For the crayfish that I study (and other aquatic animals), the types of molecules that "olfactory" and "taste" sensory systems respond to are similar if not identical to each other. In addition, the types of neurons are also very similar in function and structure. Scientists often classify smell and taste in aquatic animals by their behavioral uses as both systems detect distant odor sources. (Finally, for ease of use, I will repeatedly use odor to indicate any type of molecule that activate the chemical senses.)

2.1 Stay for a While

The lingering of a signal (odor molecules) is conspicuously absent in visual and auditory signals. As I give the command "come" to Cedric, for he is lingering too long at an oak tree, the sound wave leaves my mouth and travels for a good distance. He hears the command and politely responds by ignoring me. The moment that I stop producing the sound "come," that signal is no longer present from the environment. (To be truthful, the sound is present as it moves away from me at roughly 1000 feet/second, but after a couple of seconds, that stimulus is gone from the environment). Although Cedric hears the command, he is intensely interested in the oak tree. The base of the oak tree has layer upon layer of the odor of previous dogs that have walked by and Cedric is reading this guest book with his nose. Each of those layers is a "name" that says "Hey, my name is Rover and this is my territory."

In Chap. 1, I described the red shouldered display of the blackbird as he is setting up his territory. With the full sun reflecting off of the red patches, his signal is quite conspicuous. As soon as the sun fades or the bird flies away, the signal vanishes. Given the tremendous speed of light, visual cues are generated and sent instantaneously. This characteristic is a tremendous advantageous when speed of transmission is a prime factor for evolutionary selection like when a hawk is bearing down on the blackbird. The problem with visual signals is that they disappear just as quickly as they appear. Any birder is keenly aware of this problem with visual signals as they attempt to catch a glimpse of their favorite woodpecker among the trees of a forest.

The ability of chemical signals to stay present in the environment even when the original source has moved on allows organisms to use these signals in ways that sight and sounds cannot be used. Scent marking, nest mate recognition, infant–mother connections, and odor memories are just a few unique examples found in chemical signals. Eventually, the odor molecules deposited by the previous dog will slowly diffuse away and the smell (Rover's name) will no longer be evident. That is unless Rover feels the need to lay down another signal at this particular tree. The long lasting characteristic of chemical signals is particularly evident for those odors we do not particularly like.

A previous four legged companion of mine, a playful black lab named Loki, had a particular habit of chasing and toying with neighborhood skunks. Unfortunately for me, Loki had a difficult time learning to avoid these animals. Time and time again, she would chase them down only to receive a dose of their "perfume." The dire consequences of her escapades would become quite evident as soon as we would let her in from our backyard. Immediately, our house would fill with the distinctive calling card of the skunk; the nasty little molecules (a class of compounds called thiols) would spread through our house. We would immediately dash into action, washing her down with special soap to remove the offending odor. Although the skunk was never in our house and had most likely departed our yard, its presence was certainly felt as the signature perfume permeated our house. Even weeks later, Loki would occasionally smell "skunky" as rain or humidity would release the skunk's second and more secretive weapon: a second class of chemicals, thioacetates, remain attached to Loki's fur until moisture converts them to the more pungent thiols. This reactivated calling card provides an evolutionary advantage of reminding the would-be predator of what happens when it attempts to attack a skunk.

2.2 Any Which Way the Wind Blows

One last tug on the harness and a little more forceful "come" alert Cedric that it is time to continue our walk. His ears pick up, tail curves over, and off we go down the sidewalk again. Although he does not stop his slow trot, Cedric is still connected to the world of odors around us. With a far less sensitive nose, I am admiring the colors of the flowers and trees on my short walk and listening intently for any song bird that may entertain me. In the meantime, Cedric is quickly sniffing to the left, to the right, up in the air, and along the sidewalk. We are all most likely familiar with this sniff as it is a characteristic of most if not all vertebrate animals. Whether we are sampling laundry to see if it is clean, checking the freshness of an unidentified food container in the refrigerator, smelling the intoxicating fragrance of a baby's head, or partaking of the wondrous odors of the fresh cup of morning coffee, we sniff in order to interact with our odorous world.

The behavior of sniffing is a requirement for the sense of smell and points to a second unique feature of the chemical senses. Chemicals need assistance to get from point A to point B, and that help involves movement of air or water. As Cedric and I continue our walk, we approach an intersection with a stop sign. The characteristic red of the sign needs no help from an outside agent to reach my eyes. Inherent to the color red, and to all light, are properties that move that signal through the world to my eye. For all practical purposes, the light leaves the sign and is instantaneously at the receptors in my retina. If the sign were to change colors, say blue for instance, the color would travel just as fast, but because of the shorter wavelength there is a greater potential for the light to be reflected off of small particles in the air. For visual signals, the color of the light not only conveys meaning (red equals stop/halt) it also impacts how the signal moves through the environment. The sky is blue because the shorter wavelength is scattered by particles in the atmosphere. In a similar fashion, sound has inherent properties that move that signal through the environment.

Fig. 2.1 Odor plume

This situation is very different for chemicals. We can perform a little thought experiment to demonstrate this difference. Let us say, I take a handful of those skunk's thiols, place them on a table, and remove any air movement from a room. Next I invite you into the room and ask if you can notice anything. You would not be able to smell the skunk odor at all even though those chemicals are present within the room. The moment I placed a small fan behind the chemicals, you would run from the room and probably not accept any more invitations from me. Whether those chemicals are the scent of a skunk, red roses, or cherry pies, you would be essentially blind to the odors without a little air movement to aid the delivery of those molecules to your nose. When Cedric sniffs, he performs the same trick as that small fan in the room. During the sniff, he draws those molecules from the environment and delivers them to his nasal passage in order to determine what he is smelling.

On a scale significantly larger than Cedric's nose or the room with the fan, the wind does the work of spreading, transporting, or dispersing odorants. The pleasant smells of fresh cut hay in the middle of August, a set of barbeque ribs on a grill, or the sweet fragrance of a blossoming lilac bush would be hidden from us without at least a gentle breeze blowing toward us. We call these wind delivered chemicals an odor plume (Fig. 2.1). The same phenomenon occurs within aquatic habitats. Water movement is necessary to deliver signals to their waiting recipients.

The interaction of odor molecules with the wind or water that moves them gives rise to the sensation we call olfaction. Unlike the red stop sign or my command "come," odors are almost always intermittent. The red from the stop sign is constant and as long as I produce a sound, the voice command is there. Imagine walking through an English Garden in full bloom. During your pleasant stroll among the vibrant colors, you would receive periodic whiffs of the various flowers even though the flowers are constantly emitting their calls to bees. If we could visualize the movement of odors through our imaginary garden, we would see a mixture of puffs shaped like little mini-clouds with fine tendrils of odor between those puffs. The scene would be similar to the white wisps of steam emanating from the chimney of a steam plant. In our garden, each flower, each stem, and blade of grass is a tiny

Fig. 2.2 English garden plumes

Fig. 2.3 Odor landscape

chimney churning out their unique blend of chemicals. As we walk along our path between the flowers, each of these odor plumes mix together to form the odor symphony that is the English Garden (Fig. 2.2). As bees, being quite smaller and faster than we are, fly through the garden; they are greeted with giant chimneys creating great puffs of odors.

We can scale this analogy to both larger and smaller sizes if need be. An entire farmer's field is a thousand chimneys of corn each producing their own odor plume. Next to the field could be a small wood lot and within the wood lot, every living thing (and some nonliving) is producing their own plumes. With each and every exhale, the squirrel high in the branches of the trees creates a signature plume. All of the leaves, as they respire, produce chemical upon chemical wafting through the woodlots wind. This image, writ large, is the world of odors and organisms great and small all have their unique set of odor plumes (Fig. 2.3).

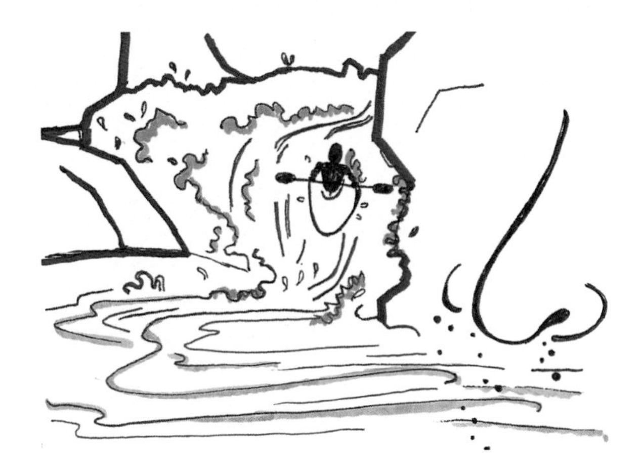

Fig. 2.4 Kayaker riding an odor plume

A third contrast can now be made between the chemical world and the visual and auditory worlds that we live in. That stop sign at the crossing is constantly on, as long as there is light shining on it. I do not have to 'locate' it because there is a direct path from the red light emanating from the sign to my eyes. My command to Cedric travels directly from my mouth to his ears and by comparing the difference in arrival time between his right and left ears, he looks directly at me knowing that I am the source of the sound. The odor world is directly impacted by the movement of air and water without which we would be odor blind.

On our human size and time scale, air and water move through their respective environments in what is called turbulent motion. Within turbulence, streams of water or air crisscross in a chaotic fashion. Think of the last bumpy flight you might have taken or imagine pictures of kayakers racing down the Colorado River (Fig. 2.4). Traveling through this turbulence, the plane bounces up, down, left, and right in a way that is impossible to predict. The kayakers (as representatives of odor molecules) have little control compared to the power of the river over where they go. They are thrashed about as the river delivers them down stream. The seemingly random motions of the fluid (both air and water can be treated as fluids) causes the puffs and tendrils of odors to be torn and shredded as they move from the source of the odor to our nose. What we perceive when we detect a turbulent odor plume is an interesting mixture of odors that vary wildly in intensity. In one moment, the odors appear to be quite intense and close, whereas in the next moment, we sniff empty air with no odor. Locating the source of these odors is complex and challenging task because of the intermittent and varying nature of the signal as it moves through the environment. Imagine trying to find a rabbit dyed pink among a warren of rabbits. For odors, this simple task would be akin to attempting to locate that same pink rabbit let lose in a Las Vegas night club while someone was flashing the lights on and off.

Cedric is fully at home in this chaotic world of swirling odors. He can navigate these plumes and locate their sources with tremendous precision honed through natural selection. We, on the other hand, are lost in the whirls and whirls of odor parcels moving somewhat randomly around our nose. As we sniff, these odors seem to randomly appear and vanish as we move through our environment. As I glance around and take in the visual landscape during our walk, I can easily reconstruct the visual image in my mind. Without too much difficultly, I could recall the spatial relationship of the previous stop sign, the side walk, and all of the houses we are passing. I can create a map of these different locations and even with my eyes closed, I could point, in a rough direction, to anything queried about. I can do this because the visual landscape is relatively static and what that means is the images within the landscape (the houses, stop signs, and sidewalks) have a set location that is constant through space and time. This is not necessarily true if the visual sources themselves are moving such as birds, cars, or clouds.

The odor landscape is dynamic regardless of whether the chemical sources are still or moving. Quick shifts in the wind (or water) cause the odor plumes to shift dramatically in both space and time. A sudden gust of wind and odors can disappear completely. Another shift and the plume work its way back to your nose. This dynamic landscape is a fundamental aspect of the world of odors. The sensory landscape of odors is a constant shift in mixtures and intensities that for the most part we are blithely unaware of. This landscape can be quite frustrating to us as we try to find the source of a sour smell inside of our house or some other offending odor.

Despite this unstable stimulus environment, many animals have evolved mechanisms that allow them to thrive by using chemical signals. In subsequent chapters, I will cover some of the more interesting ways in which a wide variety of organisms use chemical signals to locate their sources and ways in which this dynamic landscape influences our own behavior.

2.3 By Any Other Name Would Smell as Sweet

After entering the University campus, we turn left, head down a small slope in the sidewalk, and approach my office in the biology building. I get a sense that Cedric knows we are close to my office because he picks up his pace ever so slightly. A very stoic dog, I still perceive a subtle shift in his body posture which, to me, signals an increase in his excitement. This shift is probably caused by two things that await him when we enter the building. First, and probably far more important, is a box of dog treats. This is his first reward for making the trek into work with me and for not sampling the flowers too often on our walk. The second treat is the host of individuals that want to pet, hold, or say hello to him as we make our way to my office door. This is jokingly known as puppy therapy as Cedric is more than obliging in letting people hold him or pet him. Timid as well as stoic, Cedric is cautious even with people that have known him since the day we brought him home.

Because of his timid nature, everyone walks up to him with their hand in front of them to allow him to smell them in order to recognize them. Each of my students has a unique smell which would identify them as Sara, Thom, Maryam, or Ana. In nature, this signature mixture of chemicals would be a combination of chemicals derived from their diet and a set derived from their daily experiences. For example of these unique odors, a beef loving farmer would have a distinct odor from a pork eating botanist. Today, we tend to mask these odors with scented hand soap, shampoo, perfume, and cologne. Even if we forgo the daily cleansing and covering ritual, we would have a singular signature smell that would allow a blindfolded Cedric to determine your identity.

In this sniffing and greeting behavior, we can find the fourth distinction that chemical signals have when compared to visual or auditory stimuli. When we scientists consider sensory stimuli, we often categorize the stimuli based on important features. These features are considered important because they carry information that allows the receiver to make sense of what the signal is conveying. With visual signals, color is an important feature. The stop sign we passed on our walk is red. If we were to change that color to yellow or green, we would be conveying very different meanings. My command of "come" has a certain frequency (or suite of frequencies) that is characteristic of my voice and the command issued. Interestingly, both of these signal features (color for light and frequency for sound) reside along a linear scale, and their position on that scale influences, in part, how those signals move through the environment. By linear, I mean a systematic organization that moves in one dimension. Looking at a rainbow or the cover of the Pink Floyd album Dark Side of the Moon, one can name the colors in descending order of wavelength size by remembering the mnemonic device Roy G. Biv (red, orange, yellow, green, blue, indigo, violet). Sounds, too, have a linear relationship that influences their movement through the environment. Comparing the differences between the signals in these three sensory channels, light, sound, and chemical, helps to illuminate how humans use information from these senses in various ways.

We perceive rainbows because different colors of light have different wavelengths, and these wavelengths influence how these colors move through the environment. Blue color has a much shorter wavelength than red light. If we wanted to extend the spectrum of light more fully, ultraviolet light, often invisible to us, has even shorter wavelengths than blue light. At the other end of the spectrum, there is heat which has very long wavelengths that are detectable with special cameras and special sensory structures such as pit organs in snakes. Taken in totality, the electromagnetic spectrum stretches from ultrashort gamma rays to the exceedingly long wavelength radio waves. The morning sky and large bodies of water often appear blue because the short wavelength of this light gets scattered easily. What this means is as this light travels from the sun to our eyes, tiny particles in the atmosphere redirect the blue light to scatter it throughout the sky. At the other end of the spectrum sits red light. Red light, with its long wavelengths, does not get scattered as easily as blue light, but does get absorbed rather quickly. The point of this little detour into light and sound is to explain that these patterns of absorbance and scattering are perfectly predictable based on physical theories. Within the world of odors, there is no simple relationship between signal (the color of a red stop sign or the sound of a tenor aria) and its movement through the environment. Thus, any kind of organization of signals is quite messy.

This messiness is readily apparent when there is any attempt to classify the psychology of odors. Naming and classifying the world of odors is called determining odor qualities. The concept of odor quality is quite different than measuring the wavelength of red light (620–750 nm) or the frequency of my command to "come" (somewhere between 85 and 155 Hz for my baritone voice). Both of these measures are independent of any receiver psychology which includes the neural architecture that underlies the reception of sensory signals as well as the ability to discriminate and remember signals. Scientists have instruments that will measure these values and these measurements should be identical whether one is 15 or 80, French or German, or male or female. Odor quality is a very different concept.

The field of psychophysics is a branch of study designed to elucidate the relationship between stimuli and human perception and sensation of those stimuli. Psychophysics has developed a number of terms to describe odor qualities such as earthy, floral, minty, spicy, and fishy. As standard as these terms have become within the psychophysics literature, fishy or spicy is still quite subjective compared to 120 Hz or 710 nm for auditory and visual signals and is very dependent upon the person being asked to identify the smell.

Having spent 4 years on Cape Cod during my graduate school time, my wife and I absolutely fell in love with fresh seafood. Some 25 years later, the sweet smell of a freshly cooked lobster or the aroma of mussels simmering away on the stove is joy for us, and in many ways, the odors evoke powerful memories of a scientific youth spent walking the hall of the Marine Biological Laboratory. Thus, the odor quality of seafood has an additional element of pleasantness added into that fishy or lobster smell. Unfortunately, our children do not feel the same about these odors and are often eager to leave the house when a seafood dinner is planned. This is the subjective nature of aromas and many researchers have spent years attempting to tie together the perceptual quality of an odor to something about the structure of the molecule that evokes that quality.

Chemical compounds that stimulate our nose come in a large variety of shapes and sizes. Organic compounds, by definition, have a carbon atom as the backbone of the compound. The backbone consists of a series of carbon atoms hooked together with either a single or double bond. The backbone can be as short as in a single carbon atom in methanol or methane to extremely long chains of carbon atoms found in some of the toxins in marine Dinoflagettes which have over 300,000 carbon atom chains.

In addition to this backbone, carbon chains can have different configurations of atoms somewhere along that backbone which gives the compound its unique smell. This part of the molecule is defined as its functional group and are important from a chemical point of view because the exact structure determines the majority of the molecules' chemical properties (volatility, boiling point, freezing point, etc.) and more importantly for this book, the functional group provides the unique odor quality to the molecule. Any change in the molecular configuration or shape of the chemical compound has the potential to change our perception. The functional group and length of the carbon backbone are the two main elements that determine the odor quality of any particular chemical compound (Fig. 2.5).

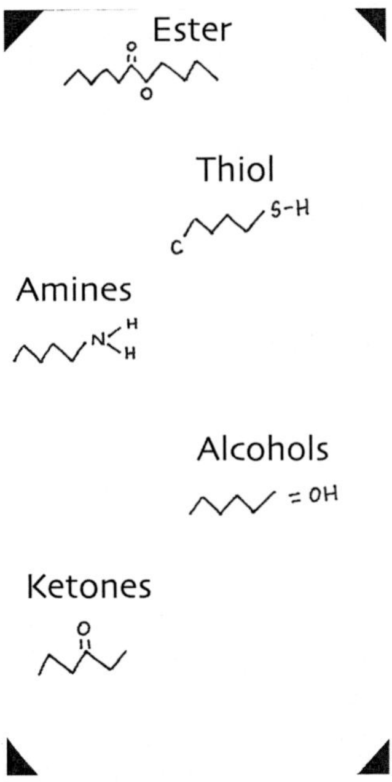

Fig. 2.5 Functional groups of chemicals

Ester

Thiol

Amines

Alcohols

Ketones

Amines (nitrogen connected to two hydrogen atoms) tend to be some of the more offensive odors. If we were to construct a backbone of four carbons and add the amine functional group to both ends, we should have produced the aptly named putrescine which smells like rotting flesh. Add another carbon to the backbone? We have the sister compound named cadaverine which too smells like rotting flesh. As an odor group, amines are relatively easy to categorize. They are somewhat offensive to our nose and can be described as fishy or repulsive. Different functional groups will have wildly varying odor qualities and to explore the totality of human odor perception is really beyond the scope of this book. Yet, we can take a mental stroll through the department store of nature. Instead of an annoying sales assistant ready to ambush you with unannounced spritzes of the latest perfume, Mother Nature herself is here to simply point to different bottles for our nasal pleasure. After quickly passing by the amine counter (which really smelled like a fish market), the wonderful makers of the alcohol odors are ahead on the left. Alcohols are organic compounds with an oxygen and hydrogen as their functional groups. As with other chemicals, alcohols vary greatly in their odor quality. Simple alcohols, such as methanol (one carbon) and ethanol (two carbons), are probably quite familiar to us as fuel smells. Adding carbons to these molecules does something amazing to the odors. Instead of smelling like fuel or drinking alcohol, the odor becomes fruity and light when the number of carbon atoms in the backbone hits six and seven (hexanol and heptanol, respectively). Adding even more carbon atoms changes the perception of the odor and the chemical smells floral or sweet. At the far end of the alcohol counter is a small little bottle with a hexagonal shape on the label. (In addition to the branching of carbon atoms above, some organic compounds have rings of carbon atoms attached to the functional group.) Upon lifting the glass stopper on the bottle and bringing our nose carefully to the end of the stopper, a cool sensation passes over us and goose bumps appear on our arms. This bottle contains menthol which not only activates our sense of smell but also triggers that little known third chemical sense called the trigeminal sense. (More on this sense later in the chapters). This odor is often described as refreshing and stimulating. Having sniffed enough of the enormous variety of alcohol-based fragrances, the next perfume counter awaits our arrival.

If we close our eyes as we approach this set of fragrances, we could imagine that we are at a fruit stand rather than nature's department store. This counter contains a number of different esters from around the world. Esters are characterized by the presence of two oxygen atoms often in the middle of the carbon backbone. One of the oxygen atoms has a double bond attached to the carbon and the other oxygen is sandwiched between two carbon chains. These chemicals are often found within fruits and flowers and provide that sweet smell often associated with the ripening of fruit. This perfume counter full of color-coded bottles is just waiting to pleasure our nose. The closest bottle is a bright yellow with a few speckled dots that appear randomly sprinkled across the label. Lifting the stopper, our nose is greeted with the very familiar smell of bananas. The chemical stimulating our senses is called isoamyl acetate and is often used as banana flavoring in many foods. A large and diverse family of chemicals, these esters are found in apples (butyl acetate), cherries (geranyl butyrate), and even honey (methyl phenyl acetate). As we sample bottle after bottle, the subtly different chemical compounds all bring to mind fresh fruit ready to be eaten. Our noses are in a veritable heaven when we notice two separate bottles at the end of the counter. The first bottle is plain and white, while the second bottle has no label and is simply clear. Intrigued by the mysterious colors, we carefully sniff the white bottle and our minds are flooded with images of elementary school craft days. The chemical (methyl acetate) is the main fragrance in white glue. After clearing our noses of this odor, the clear bottle draws our attention. Upon uncapping this bottle, the pungent odor of nail polish remover is released into the air. This compound is yet another form of an ester (ethyl acetate) and is distinctly different than the fruity odors sampled previously.

We could very well continue through the mental construct of the store's perfume section, but we would be there for quite some time. In addition, to the amines, alcohols, and esters, there are thousands of other compounds like ketones, acids, phenols, and other aromatic organic chemicals. None of these fit into a neat and tidy organizational structure like sound or sight. In some ways, the qualitative nature of chemical signals (i.e., our reliance on a person's perception of an odor quality) has hampered the scientific progress in understanding the human nature of this sense. In other ways, the exciting unknown nature of new aromas has been a boon to those interested in developing the latest new taste or odor. In particular, a lot of research (both scientific and not so scientific) has gone into the search for pheromones.

2.4 The Scent of Love

Probably the word most associated with the sense of smell and the most misunderstood is the word pheromone. In many of my conversations with people on this topic, the word pheromone conjures up many different images often revolving around sex or reproduction. The most common description given to me during these conversations center around an aroma that is so powerful and overwhelming that a mere sniff turns the receiver into a behavioral zombie unable to think of anything but sex. Pheromones are thought to be nature's aphrodisiacs on steroids. A mere drop of the compound will draw in males (of the same species) in from miles around in a frenzied state. Like most common interpretations of scientific concepts, there is only a kernel of truth to this description. The most common use of the term pheromone, particular in early instances, was associated within mating in animals. In reality, the definition of pheromone is simply a chemical signal released into the environment that evokes a behavioral response in organisms of the same species. The first instance of the identification of a sex pheromone comes from the silk moth.

Female moths, of the species Bombyx mori, produce a compound with the explosive moniker bombykol (taken from the Bombyx genus name). Discovered by Adolf Butenandt in 1959; Butenandt was a German biochemist who won a Nobel Prize at age 36 for his work on sex hormones. Given the economic importance of the silk industry at the time, Butenandt and others had ample access to both the males and the females of this species. Upon smelling the pheromones, male moths initiate a series of behavioral patterns that appear to be unstoppable. They begin to fan their wings as if warming up for an intense flight. Once ready, they launch themselves into the air and by a mixture of cross wind search and upwind flight for intense spots of aerial sex pheromone, the males make their way to distant "calling" females. Interestingly, we use the auditory term "calling" to describe the female's act of pumping out pheromones to draw the males in. Typically, the females will climb some relatively tall structure (tall for them) and initiate their pheromone release. Once the males are within vicinity of the female (same tree or branch), they will land and begin a local search for their prospective mate. At a glance, the behavioral acts of the male silk moths do appear to be "involuntary" as if the female's pheromone has put the male into a sex-starved zombie state. It doesn't do this to males, but to understand why there is this belief, a brief history of the word pheromone is necessary.

A quick search on the internet or through any number of magazines will quickly show how the concept of pheromone has been stretched beyond recognition and into the realm of magic love potions. The first broad categories of searches will produce a laundry list of sites where one can purchase small amounts of special chemicals that will give the wearer unstoppable sexuality, power, and control. Sold only in very small quantities, these ultrapowerful scents are guaranteed to give the wearer what they desire. Obviously, the concept behind these scams is that the recipient of the pheromone scent can't help fall in love or be awed by the power of the wearer. Some sites even claim that the pheromone is odorless and yet still functions to elicit the appropriate behavior. Another product (aimed at men) will give the wearer trust and respect. As the site claims, both men and women will "give you their attention and respect, and will want to take your lead. You'll find more people than ever 'bending over backwards'". Finally, a new trend within the dating circles is something called pheromone parties. Days before the actual party, attendees wear identical shirts (blue for boys and pink for girls) for 3 days. The idea is that the essential smell of the individual will be trapped in the t-shirt. Party goers arrive at the host's house with t-shirt in a plastic bag with a number attached to the bag. Over drinks and snacks, participants walk around and sniff shirts in hopes of finding their one and only soul mate. Of course, that soul mate has to be completely compatible as determined by the nose. The "science" behind this concept is that the personal pheromones will quickly trigger a set of biochemical pathways associated with love. So, instead of love at first sight, it is love at first sniff.

The term pheromone originated around 1959 and was constructed from two separate words. The first part, phero, is derived from the Greek word, pherein, which means to transport and the second part, mone, comes from a shortening of the word hormone, which means to stimulate. Several researchers at that time, including another Nobel winning scientist Karl von Frisch, developed the terms to describe a set of chemical signals transported outside of the body in order to elicit a behavioral reaction in the receiving organism. During this scientific period, the concept of innate behaviors was quite prevalent. Innate behaviors, at this time, were thought to be a set of behaviors that animals were born with and that were identical every time the behavior was enacted. Thus, the concept of pheromone, releasing a behavioral reaction, and innate behavior, something not learned or altered, were unfortunately brought together to produce the concept of pheromones presented above.

In reality, the sex-starved, love struck, soul mate zombie producing pheromone doesn't exist, but numerous pheromones that produce different behavioral reactions in the receiver do indeed exist. Pheromones are a sub-class of chemical signals, and the term is reserved for those chemical signals that have evolved to produce reactions with conspecifics (same species). Pheromones, then, are released by one organism in order to produce an expected behavior in another one of the same species. To differentiate between the terms chemical signal and pheromone, we can return to the story of my black lab and skunk. The skunk produces the wonderfully odorous set of chemicals evolved to ward off or deter predators. These cues are interspecific signals or signals where the intended recipient is a different species: my black lab. Antipredator cues or compounds within plants that deter herbivory are just such types of compounds. Conversely, pheromones are those chemicals that are intended to send information to organisms of the same species. There is a host of chemicals that fall within this category.

Another potential misconception with regard to pheromones is that these are a single chemical. One can think of an ancient search for the holy grail of chemicals. Among the popular concepts of pheromones is there is a single chemical (or grail) that will produce the desired behavior. One drop of the sacred compound and the "victim" is hopelessly under control of the wearer. Pheromones can be mixtures of compounds, although that mixture is tightly controlled by the biochemical production in the perfume factory of the body. Often there is a predominant chemical within a pheromone mixture, but the mixture is needed to really call the signal a pheromone. A fuller discussion of pheromones, in particular regarding sex pheromones, awaits in Chaps. 8 and 9.

So if pheromones are used for things beyond reproduction, what ranges of behavioral situations are pheromones found? Pheromones can be grouped in regard to their behavioral function (almost similar to the chapter headings of this book). First and foremost are the commonly recognized sex pheromones. This class of compounds are used to send information about the sex of the sender, the age of the sender (is the sender of reproductive age?), and the sexual state of the sender (is the sender in a reproductively receptive status?). Many primates, other mammals, and invertebrates use pheromones within all of these contexts. Surprisingly, birds are a major grouping of animals where chemical signals, in general, and pheromones, in particular, are relatively unknown. Pheromones are also used to send warning signals associated with predatory events or invasions from rival neighbors. These pheromones are called alarm signals. Aquatic crustaceans produce alarm signals when attacked by predatory fish that causes other crustaceans in the area to quickly find shelter. Aphids produce a chemical substance when attacked that causes other aphids to quickly disperse. Trail pheromones are used by a number of ant species to find their way to and from food resources. Finally, a subset of pheromones are called primer pheromones. Primer pheromones aren't necessarily "smelled" as other pheromones (more on this in Chap. 8), but cause significant changes in the hormones of the receiver. These pheromones "prime" the receiver's physiology for certain behaviors such as copulation. One of the best examples of priming pheromones is found in mice where female rats grow to sexual maturity faster in the presence of adult male odor. The male odor primes the female rat's physiology for sexual maturity and potential mating.

2.5 Feeling with Our Nose

During my adolescent years, my whole family was involved in our church choir. The church's music director was also the band director in middle school band where I played the baritone. Because of these overlaps in musical interests and the close relationship that the music director had with me, our two families became close friends. In fact, my parents still use one of her sons as their dentist. Through this deep musical friendship, my family was invited to visit the choir director and her family at their summer cabin in northern Minnesota.

As evidenced by my ultimate career decision as something of an aquatic field ecologist, I love nature and spending time at a cabin surrounded by woods, lakes, and streams seemed like heaven to me. Thinking back to those couple of summers visiting their cabin, I can picture the cabin in my mind's eye as if it were yesterday. It was a typical wooden cabin sitting about 30 yards from the shores of Shagawa Lake in Ely, Minnesota. The cabin consisted of a large open room that functioned as both a dining room and living room. There was the obligate and for most nights necessary fireplace with a stone mantel. The main floor had a couple of bedrooms and I believe a loft bedroom upstairs. The wooden beams which were evident everywhere on the inside were all varnished a maple color. Outside the main house located halfway between the cabin and the lake was a small wood fueled sauna. Although we visited during the height of summer heat, the lake was deep and spring-fed and would remain icy cold throughout the summertime which made swimming excursions short.

These were glorious times for me, as a young teen. The surrounding woods were a joy to explore and their big blonde shaggy dog (presumably a bushy golden retriever) would often come with me. They owned a small two person handmade speedboat and we would buzz around the lake at breakneck speeds. Fish were abundant in the lake, but fishing would require me to sit still far longer than I was capable of doing. After a day of strenuous activities that included water skiing, chopping wood, and hiking, I would often walk to the pine sauna for a short relaxing and sweaty spell. After sitting and releasing any tiredness from my muscles, I would sprint down the hill, on to the dock and plunge myself headfirst into the icy cold water. Being young, naïve, and foolish, I would repeat this heating and cooling until I got called in for a substantial supper. What made supper particularly enjoyable for me was the presence of handmade, just out of the oven bread that usually accompanied a stew. The smell of this simple, yet hearty white bread that overwhelmed us when we entered the cabin was intoxicating. I believe that most of my meals consisted solely of large slices of that bread covered in either honey or butter. It was perfect for soaking up the beefy soups and stews that made up our lunches and dinners.

The summers of my teen years, as with most people, were impressionable times, and there was an abundance of opportunities at the cabin on the lake or frequent camping trips that would influence me. The long hikes through the woods to distant streams would certainly reinforce the young biologist in me. Running through the woods, playing in the lake, the big shaggy dog, and the sounds of a forest alive with wildlife are just some of the memories that remain with me today. Yet, the two most potent stimuli for evoking memories of those summers are the smells of the pine sauna and that of fresh-baked bread.

Even now, more than two decades later, these odors have immense power over me. One of my favorite rituals after a hard workout at the university gym is a 15 minute sit in the sauna. Upon entering the sauna, the smells of hot air, steam, and heated pine wood immediately envelope my nose. This mixture of scents is hard to describe as are most complex mixtures of odors. For me, the moist heated air of a pine or oak sauna is a unique odor. Being exhausted from my taxing workout and if the sauna is not crowded, I will lay down, close my eyes (just like my sushi ritual from Chap. 1), and breathe deeply. The influence and control of odors in our lives is quite evident during that deep breath. As soon as those first molecules of heated pine reach the olfactory cells in my nose, I am immediately transported back in time to my teens and that sauna by the lake. This is not a conscious act on my part. I am not actively searching for this memory and recalling it. The odor molecules have effectively reached inside of my memory bank and activated those long bygone days. Just as if viewing a virtual reality movie of my life, I am surrounded by my past. I can hear the gentle breeze passing through the tall pines of northern Minnesota. I am glancing out the imaginary window of that sauna and can see the lake right there at the bottom of the hill. If I didn't know any better, I feel as if I could bolt out of the door and sprint down to an icy cold refreshing lake. These visual images of that distant summer flood my mind uncontrollably.

Similarly, the smell of fresh-baked white bread also transports me through space and time to my youth and to dinnertime at the cabin in northern Minnesota. These joyful memories are uniquely tied to odors that were present during those youthful summers. The phenomenon of odor memory is very powerful and the instant transportation through space and time that odors evoke does not occur with other senses. Within the fantasy world of Harry Potter, wizards and witches instantly travel to distant places through a process called "Apparition." When one apparates, the body is pulled toward the destination. Odor memories apparate our mind in that the smell of bread or a pine sauna pulls the mind back to that place in time. We all have formative odor memories that are the most vivid memories that our mind can produce. Sometimes, these memories are not necessarily pleasant (as in cases of food poisoning).

I often see big shaggy dogs during my runs or walk through town, yet none of these beautiful beasts transport me to my youth. My current summers are spent in northern Michigan performing research in the abundant streams of that area. During my treks to distant field locations, I often pass by cabins that look remarkably similar to that one in northern Minnesota. As I view these places, I am reminded of those distant summers, but my mind and body are still located in northern Michigan in the present time. As strange as it may seem, the sights and sounds of a big shaggy golden retriever, entering a log cabin, or walking through large pine forests evoke wistful thoughts of my youth, but these wistful thoughts are mere shadows when compared to the memories induced by the specific smells.

2.6 I Think, Therefore I Am

Throughout this book, I have and will repeatedly mention that humans are acutely unaware of the prevalence of chemical signals in nature. I have alluded to the idea that humans are essentially blind to the use of chemical signals in nature, and yet, we have these powerful odor memories. These two ideas are seemingly incongruent, being blind on one hand and powerfully moved on the other, and require some explanation. That explanation comes from an understanding of how odors and other senses are encoded, processed, and perceived by our minds.

Visual and auditory stimuli can be grouped together as highly processed senses. This means that, in common terms, we consciously think about the majority of these stimuli before we react to them. Imagine that part of your morning ritual involves turning on the morning news while you get ready for the day. Even if the stories about the latest world crisis are in the background, the sound waves travel from your ear through the cochlea on to the auditory cortex. Here, in the cortex, is where the brain processes the information in order for you to identify words, sentences, and extract information. If you turn the TV on to watch the morning weather, the visual signals are sent through the retina through the optic nerve to the visual cortex. In the visual cortex, the images are processed in a similar fashion as the sound waves and your mind thinks about what the eyes are seeing.

Interestingly, olfactory signals, within the mammalian brain, are handled very differently. Behavioral and mental responses to chemical signals have more of an emotional response than the highly thought-out response visual and auditory responses. The basis for this difference can be seen in how our senses send information to our brain. The mammalian brain is a very complex and intricate structure. There are numerous structures and substructures that are designed to deal with very specific types of tasks. There is an area for processing sounds, an area for processing sights, an area for movement, and so on for the many different tasks that our brain performs. In a general and simplistic view and for the purposes of this book, the brain can be divided into three broad sections: the spinal cord, the brain stem, and cerebrum (Fig. 2.6). For the purpose of this discussion, we can focus in on the middle of the cerebrum onto the areas called the amygdala, the thalamus, and the hypothalamus. The hypothalamus and thalamus, often called the limbic system, which tightly control the release of hormones into our body and the amygdala is often called the emotion seat of the brain. So, this seat of the emotions and the visceral responses that our body produces to intense stimuli is directly stimulated by the olfactory bulb. The cerebral hemisphere, and in particular the outer layer called the cortex, is the upper and outer areas of the brain that make us who we are. It is believed, with good evidence, that the cerebral hemisphere of our brain is that area that makes us aware of our surroundings and ourselves. It is where we think, where we construct abstract ideas, and it is this area of the brain that makes us conscious. The visual and auditory stimuli of our world are shipped off to the thinking areas of the brain, and the olfactory stimuli pass through the emotional centers first.

I can provide an illustrated example of how our brain processes information by taking a short trip to my garden. My gardening thumb is more brown than green, but I do make the effort to have some roses in my garden. One of my favorite roses is a lavender rose called “Angel Face.” This rose is a light purple in color, but has an excellent and powerful fragrance associated with the blooms. Now, when I see a blooming angel face, the visual image is detected by the retina in my eye. This information is then passed along after a single stop to the visual cortex of the brain. Initially by passing the emotional centers of our brain, the visual images are sent to the “thinking” centers of our brain for digestion. Among other things, I see the edges of each petal. I can recognize the lavender color as different from the deep green leaves and the blazing red rose bush next to it. There is a small metal tag that labels this rose and I see letters, and recognize the words “Angel Face.” Only after the “cognitive” recognition of these signals is this information passed on to other areas of the brain. After a few seconds, a pesky Japanese beetle lights upon the rose. In a similar fashion as the visual signals, I recognize the slight buzz of the beetle’s wings as it flies in. The sound waves from the beetle’s wings stimulate the hair cells

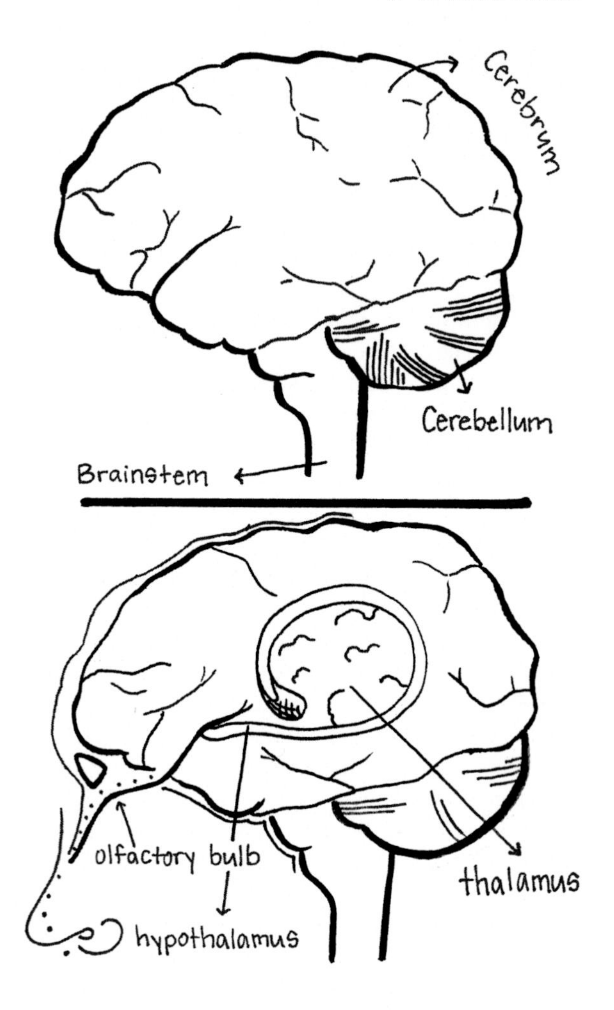

Fig. 2.6 Mammalian brain structures

in my inner ear. Once there, the information is filtered according to its frequency and is quickly passed onto my auditory cortex. Again, this information is digested and filtered so that I can recognize the sound of a flying beetle from the sound of a flying bee or bird. From this processing, I can estimate the location of the flying beetle and can distinguish this singular sound from the other springtime sounds present in my little garden. Once the buzzing is recognized and filtered, the information is passed on to the other parts of my brain.

In stark contrast to the highly cognitive processing of sights and sounds stands the wonderful fragrance that is emanating from the rose. Molecules of the rose's perfume are lifted off of the petals and are brought into my nose through a deep sniff. Once these molecules contact the receptors in my nose, neural information is generated and sent along the pathway to my brain. The first stop, the piriform complex, is a branching way station for the information. Here, the initial information about the rose's fragrance is split into two. Half of it is sent first through the thalamus and then onto the cortex. Here, like the sight of the rose and the sound of the beetle, the information is processed and "recognized" as a rose, beetle, or fragrance. The other half of the information is sent to the hypothalamus and to the limbic system. This is the noncognitive or "emotional" part of our brain. This is where aggression, serenity, love, and all range of emotions are created. Awareness, recognition, and all other "higher" brain functions are alien to this place.

What this rose example illustrates is that sights and sounds are highly "processed" pieces of information. These are sent to those areas of the brain that give us our cognitive abilities. Whereas, half of the olfactory information is sent for processing, but the other half is sent directly to the emotional centers of the brain. From the emotional center of our brain, the intense odor memories are being released and those memories flood our cognitive mind. These memoires and emotions are triggered at the same time or even before we begin to "recognize" what the smell is. Even the words commonly used to describe the impact that odor memories have upon us illustrate these differences. Words like flood and envelop are used to describe the action that odor memories enact on our mental state. Words like recall and remember are used for auditory and visual memories.

2.7 An Alien World

Cedric and I have finally made our way into my office. The usual routine at this point starts with me removing his harness and setting it down on my bookshelf. Cedric is pretty patient about letting me set my bag down and remove his harness because he knows that a delicious treat awaits his good behavior. I reach for the box and the rattling sound, which I enhance by raking my fingers through numerous treats, alerts him that his reward is at hand. Once he consumes his treat, he pads his way over to a small bed that sits to the left of my chair and computer desk. The bed is located under a section of my desk and is surrounded by a small two drawer filing cabinet on one side and the desk leg on the other side. Shiba Inus have a strong resemblance to foxes or dingos, so I can imagine that this is Cedric's version of his own den. After surveying his chew toys, he finds a spot to lay down and curls up in the shape of a gibbous moon. For the next few hours, I can get some work done and the only movement I can detect from Cedric is a couple of perked up ears tracking various sounds from around my office and the outside hallway.

At this point, Cedric is keenly aware of three different "scapes" that surround him. Those triangularly pointed ears that are roving back and forth like mini-radars are sampling the soundscape of my office. The dynamic soundscape is composed of students periodically walking in to have conversations with me, my choice of music for the day, and anyone passing my office talking on their phones, moving carts of equipment, or otherwise engaged outside of my office. Periodically, when an unexpected loud sound erupts, as when a student drops a textbook, his head will pop up and both ears instantly focus on the door of the office. If no further distressing sound arises, he lowers his head to a comfortable position on his bed, but those ears are still scanning the environment. The soundscape can be thought of as the sum total of all of the sounds, their spatial location, and the temporal dynamics of the appearance and disappearance of each of the sounds. If we were to visualize this soundscape, it would consist of waves of sound emanating from my speakers, peoples' steps, and from the mouths of students during meetings. One could think of the outward moving ripples of waves as a rock is dropped into a puddle.

The second "scape" that Cedric is aware of is the visual landscape. The visual landscape encompasses me, my desk, and the mini-fridge that stores my caffeine needs, my bookshelf, and any students or faculty that happen to walk into the office. Each of these objects creates a unique visual image that Cedric combines into a coherent and singular image. Similar to the soundscape, the landscape is dynamic in time and space. As people enter my office and move to the available chairs, the visual field is changing which is tracked by Cedric's mind. His cognitive analysis of these images determines whether he needs to walk forward because there is a treat in my hand or to stay in his "den" because someone unfamiliar is approaching.

The third and final "scape" has been described repeatedly in this chapter, and I shall call this "scape" the aromascape. Like the soundscape and landscape discussed above, each of the living beings as well as some inanimate objects produce each of their own individual aroma, and as Cedric "scans" the aromascape with his sniffs, he'll construct a mental "map" of the odors within his sensory range. I put the word map in quotes because science has yet to really understand how the spatial and temporal distribution of odors in nature are portrayed within our brain. Our understanding of the chemical senses is roughly 50 years behind our understanding of vision and hearing in part due to the difficult nature of controlling and producing aromascapes in experimental settings and in part due to our reliance on vision and hearing. Still we know that animals use odors to navigate in the world, so there has to be some type of spatial representation in their brain (and ours) of the aromascape around them. As with the soundscape, it is possible to construct a visual image of the aromascape that Cedric is sensing. Each odor source, be it student or my lunchtime sandwich, is a tiny volcano of odor that blows downwind of any air movement in my office. So, his nose "sees" 20–30 little and not so little puffs and clouds of aromas spewing forth from everything in the room. One for my can of soda, one for me, one from my trash bin, a couple from the two students who walk in, some small distant ones from people in the hallway, and so on.

This last scape, the aromascape, is the alien world that humans hardly pay attention to. The aromascape is the world that many different animals are aware of as they move deftly and respond behaviorally as easily as a mountain goat climbing the side of a steep hill. In contrast, we are bumbling our way through this world as we douse extra perfume on our bodies and walk through crowded plazas with food carts all around. In subsequent chapters, we will mentally stroll along side some of the best examples of animals using aromascapes to their advantage for specific behavioral tasks and then we will compare how the human animal is guided, controlled, and influenced to perform some of those same behaviors.

Chapter 3
Eat or Be Eaten

A Bittersweet Symphony of Food

A quiet symphony is heard in the background as I write in the corner of my favorite bakery. The music is a beautiful marriage of instruments; a blend of the soft sounds of oboes, violins, cellos, and the forcefulness of the brass instruments provides an inspiring atmosphere. If I concentrate on the music, a remarkable phenomenon occurs. Although the symphony as a whole continues, I can focus my hearing on the individual melodies of a single set of instruments within the song (Fig. 3.1). First, I concentrate on the soulful cry of the oboe, a sad melody that slowly climbs the scale. Next, I switch my focus on the French horns. This forceful staccato sound brings to mind a determined cavalry riding across a battlefield. The deep-voiced cellos sound like rolling thunder. Finally, I return to the symphony as a whole. Instead of listening to the individual voices of the instruments, I choose to integrate the individual notes of each instrument into the symphony and let the perfect harmony wash over me. Our auditory system processes stimuli in a very different way than our visual system.

The walls of the bakery are two toned: a dusty peach on top and a deeper red that starts right at the chair rail and continues on to the floor. My pumpkin muffin is a deep orange and my iced tea is a cool amber color. When I look around, I see only the single colors and, unlike the music, I cannot see the underlying individual colors that make the peach dusty or the pumpkin orange. Although I can see these many colors and their subtle shades, my eyes only have three different pigments (blue, green, and yellow). [Note that the three pigments do not correspond to the three primary colors blue, yellow, and red.] This is an interesting difference that is attributed to the overlap in the color sensitivity of the pigments and the relative distribution of pigments in our visual system. Pigments are those molecules located on the surface of the eye's photoreceptors that absorb the light and allow us to see color. As blue light enters my eye, the light is absorbed by a pigment. This pigment then changes its shape and excites the photoreceptors in my eye. These receptors pass this information on to other neurons and after a series of neuronal connections, my mind registers the color blue, all of which occur in milliseconds. I see far more colors than the basic three pigments in my eye, and that is because the different

P.A. Moore, *The Hidden Power of Smell*, DOI 10.1007/978-3-319-15651-4_3

Fig. 3.1 Symphony of odors

colors essentially excite the pigments in proportion to their representation in the color. Almost like mixing paint at your local hardware store, a dab of red, three dabs of blue, and a half of yellow create a light purple color. For a simplistic example, imagine standing on the shore of a Caribbean island and looking out at the crystal clear sea. The ocean around these islands often appears blue-green in color. When one looks at the ocean, this light (half green and half blue) excites my green-absorbing pigment and the blue-absorbing pigment equally. For our visual system, this method works quite well and allows us to perceive the subtle differences between shades of teal and sea foam. When green and blue light are mixed together, we perceive the scene as teal-colored light. Try as we might, we only see the color blue-green and we cannot perceive the original green and blue. The individual components are forever lost in the final color. This is a gestalt way of viewing colors. (Gestalt is a German word meaning shape or form and is encountered most often in the fields of neuropsychology or psychology. The concept is that the brain or mind conceives of a sensory stimulus as a whole rather than the all of the sub-elements. Think of the classic pointillist painting "A Sunday Afternoon on the Island of La Grande Jatte" by Georges Seurat. A gestalt approach to viewing the image would be to ignore the individual dots of paint and view the scene as a whole). Without years of training or highly sophisticated computer equipment, it is virtually impossible to tell which individual colors are mixed together to form the final product. Interestingly, our perception of odors and tastes is more similar to our auditory than our visual experiences.

As I entered the bakery, the aroma of the morning rolls and French roast coffee inundated my sense of smell. The bakery bombarded my nose with cinnamon, amaretto, sourdough, pumpkin, and French roast. I can sit back, taking in all the odors at once, like the sounds of a symphony, and form a "gestalt" of the bakery. Conversely, I can also analyze each of the odors and detect the individual scents. One of my favorite breakfast treats is a pumpkin muffin. As I bite into my muffin, a myriad of spices send me to paradise. Tasting beyond any individual flavors, I enjoy the totality of the muffin, the gestalt of pumpkin. With my second bite, I shift my focus instead to the individual tastes. First and foremost is the pumpkin itself.

An excellent mixture of moistness and sweetness swamps my chemical senses. Although we often speak of taste when we consume foods, the majority of the perceptions we experience while eating arrive through our sense of smell. When food is in our mouth, odor signals travel through the back of our mouth to arrive at our nose. To test this yourself try the "plug the nose" experiments discussed in Chap. 1. (Since we psychologically integrate both odor and taste sensations, I will use the term smell as indicating both of these terms for the rest of this chapter.) Hints of spices add to the pumpkin flavor. Cinnamon adds a touch of sharpness, and a gentle kick comes from ginger. Nutmeg adds an earthy tone, and the coup de grace comes from a fine dusting of powdered sugar.

Imagine that a young baker in training has the opportunity to make the morning muffins. She dutifully mixes in the sugar, flour, eggs, butter, pumpkin, and other ingredients. Right when the spices are due to be mixed into the recipe, the phone rings and she forgets to add the ginger. From a nutritional point of view, the gingerless muffin is equivalent to the other muffins. However, if I were to taste this new muffin, a different symphony of odors would find its way to my nose. I would recognize this new muffin as quite similar to the earlier version, but something would be amiss. If properly trained, I might be able to recognize that ginger was missing. Now, if my nose operated the same way as the visual system outlined above (as a gestalt only system), I would perceive the pumpkin muffin as something entirely new and different. One bite of my gingerless pumpkin muffin would send me back to the counter asking for a refund because the muffin in my hand is clearly not a pumpkin muffin. A gestalt-only view of the odorous world (as in our visual world) would tell me that this is not a pumpkin muffin—close yes, but not what I ordered.

In an evolutionary sense, why should it matter whether we can perceive pumpkin muffins as the individual components of pumpkin, nutmeg, ginger, and powdered sugar as opposed to the taste of the muffin as a whole? As human beings, we eat for both pleasure and purpose. However, for all other organisms the act of consumption serves the purpose of acquiring both needed nutrients and building blocks of life and the chemical energy needed to run the machinery necessary for life. With these two evolutionary purposes in mind, we can look at the role of smelling and tasting in a new light. Along with serving as the window to the wonder of well-cooked meals for humans, perhaps more importantly, smell provides other organisms with vital information about the nutritional quality of a meal. Our bodies utilize the sugar as a source of energy that is needed to run the daily functions of our body. My pumpkin muffin is an excellent source of sugar, probably more than my body needs. My gustatory system allows me to perceive the amount of sugar in the muffin independent from the kick of ginger or earthiness of nutmeg.

A more natural example originates from the ability to detect potentially harmful compounds, such as poisons. Poisons usually have a bitter taste. For these chemicals, our life and death rests in our ability to detect the presence of the poison independent of the other flavors in the food. How effective would our smell and taste system be if a small bit of poison could be masked with other chemicals found in food? This helps explain why the ability to perceive smells as both individual components and as a whole is a better system overall. Why and how did this system come to be?

These questions cannot be fully understood until we see how organisms use their sense of smell and taste to effectively find food or avoid predation in nature. This chapter is about how organisms have evolved chemical signals that are used in one of most basic aspects of life: eating. How do organisms locate and identify food? How do they stop from becoming a meal for some other animal? As we shall soon see, organisms have very complex methods in which chemical signals are used to communicate to each other about food or even about potential predators to protect themselves from being eaten.

3.1 The Paradox of Being Delicious and Avoiding Predation

If there is a predator, such as a fox, hawk, or even an insect, there must be a prey, be it rabbit, mouse, or oak tree. If you are the predator, this interaction can be a very positive thing, but if you are the prey, this event is another story. Within this dichotomy is the understanding of the basis for all predator–prey interactions. Predators are constantly under selective pressure to have more efficient ways of detecting, locating, and catching prey. Sharper eyes, better ears, and faster legs are just some of the ways in which predators improve. Along this same vein, prey are also constantly adapting to avoid getting caught—better camouflage, faster muscles, or more potent defenses. This has been called an "evolutionary arms race," harkening back to the Cold War era when the United States and USSR were in a race to produce the most military arms. As one side, say the predator gets faster, the prey must also pick speed or risk extinction. As the prey develops better camouflage, the predators develop sharper eyes to detect their hidden prey. Upon closer inspection of this arms race, the fittest prey, the younger and stronger, typically always have a slight advantage over the predators. As the old saying goes, "The fox is only running after its lunch, whereas the rabbit is running for its life." Meaning, that if the fox makes a mistake, the fox only goes hungry until the next rabbit appears. However, if the rabbit makes a mistake, the rabbit has made a fatal error. Some of the more obvious developments in the arms race, faster legs or sharper talons, are some of the least interesting aspects of this story. A far more interesting story lies in the chemical arms race. One of the more fascinating examples of a chemical adaptation to avoid predation occurs deep on the ocean floor.

Mollusks are a group of organisms that include snails, oysters, clams, and mussels. Each member of this group of organisms has an organ called the mantle. The mantle produces the hard calcium shell that we envision when we think of the many mollusks that find their way onto our dinner plates. The shell is an excellent example of an antipredator adaptation and confers upon clams and oysters adequate protection against the many different predators in the ocean. Although all organisms within the group called mollusks have a mantel, not all of them have a shell as protection against the harsh undersea world. Some, like the squid, have internalized their shell, while others, like the sea slugs and some of the cuttlefish, squid and octopi, have abandoned the shell altogether.

Fig. 3.2 Sea hare

Without a hard shell, a mollusk must develop other methods of protection from would-be predators. Cuttlefish use a complex system of color changing organs, called chromatophores, to camouflage themselves. Octopi have the ability to both crawl along the ocean floor or, if need be, to quickly propel themselves by creating a powerful water jet. We are all probably aware of another antipredator adaptation of the octopus: its famous ink. If harassed by a predator, the octopus releases ink into the face of the attacker in order to confuse and blind them. Under this cloak of darkness, the octopus jets away to freedom. This system is a wonderful example of an evolutionary adaptation to solve a specific problem. The ink and jet system works well for the octopus. Yet for one of its cousins, the sea hare, the jet system has a single design flaw. For this system to work, the animal must have the ability to quickly swim away during the confusion created by the ink. The poor sea hare does not have this ability. They are named not for their rabbit-like movement patterns but for their diet, which consists of seaweeds and sea lettuce, and to two sensory structures that stick up off of their head-like rabbit ears (Fig. 3.2). What if you are a sea hare, a relatively slow-moving animal confined to the two-dimensional world of the ocean floor? If you are one of these nice, soft, and delicious mollusks, why would you rid yourself of a hard shell? How do you avoid becoming a predator's lunch? In spite of the sea hare's shell-less body, there appears to be no predator that makes its living eating this sea hare. A number of benthic crustaceans, namely lobsters and crabs, have been known to enjoy a sea hare or two from time to time, and sea anemones will consume sea hare when they get the chance. However, the sea hare (*Aplysia* sp.) is not a main staple in the diet of any animal species. The answer lies in chemicals.

One species of the Californian sea hare, *Aplysia californica*, has further developed another ingenious anti-predation ink method for dealing with its predators; a mechanism that was quite unexpected. Just as its cousin the octopus, the Californian sea hare can also produce ink—in fact the sea hare produces two different types of ink. One ink is produced and secreted from the purple gland, appropriately called the purple ink, while the other secretion is produced from the opaline gland which produces a gooey, whitish substance (opaline ink). Researchers have known for years that the purple ink is released when the sea hare is harassed, similar to the octopus. Instead of producing a cloud of confusion to allow the sea hare to escape, the purple ink is smelled by the potential predators, such as a sea anemone, and produces a "bad taste" in their mouth, which seems enough to discourage further tastes. If a sea anemone happens to swallow the sea hare before the secretions are released, the anemone will promptly spit out the sea hare due to the bad taste.

The old saying "You are what you eat," illustrated in the catfish example about body odors and dietary choices in Chap. 1, is just as relevant here too. The sea hare does not produce most of the bad tasting chemicals itself; rather the chemical is acquired from the hare's preferred meal of red seaweed. Californian sea hares found in areas without red seaweed lack the reddish color common among sea hares and also lack the ability to produce purple ink. In addition, a sister species of the Californian sea hare, *Aplysia viccaria*, lacks the preference for red seaweed and lacks the distasteful purple ink, but produces small quantities of white ink (of unknown function) in its place. The production of anti-herbivory chemicals is fairly common in plants and algae. A number of organisms have developed a tolerance to these anti-herbivory chemicals—an acquired taste. Upon eating the plant or algae, these organisms sequester the compounds to use as their own anti-predation shields. If the collection and use of anti-predatory chemicals was the only mechanism of chemical signals used by the sea hare, this use of odors would hardly be unique and unexpected. But the *Aplysia* story has an interesting twist found in the opaline substance. The story begins with one of the sea hare's predators, the spiny lobster.

To fully appreciate the depth of deception and communication that occurs between *Aplysia* and the spiny lobster, a basic understanding of the olfactory abilities of the lobster is necessary. For that, we turn to the work of P.M. Johnson and Dr. Charles Derby at Georgia State University. Dr. Derby and his numerous colleagues have tried for years to understand how animals differentiate between smells. His laboratory has been instrumental in learning whether organisms smell as a gestalt or as a compilation of individual odors. The primary focus of their study has been the spiny lobster. When we think of creatures with amazing abilities of smell, the bloodhound and shark often pop into our minds. Stories abound about the tracking abilities of bloodhounds and most of us have heard of the shark's ability to detect a drop of blood in a million liters of seawater. These examples are unimpressive when compared to the lobster's capabilities. Instead of a single nose, lobsters have at least 13 pairs of appendages sensitive to aquatic odors. Dogs and sharks have millions of receptors in their nose, while lobsters have millions of receptors on a single appendage. Lobsters can learn and track odors just as well as bloodhounds, and there are several studies showing that lobsters can detect a drop of odor in a million times a million liters of seawater.

Let us return to our sea hare and spiny lobster story. The sea hare certainly could not out run the lobster, so using an ink cloud for confusion would not work. With so many highly sensitive chemoreceptors, one would think that the lobster would succumb to the sea hare's distasteful purple ink, but the sea hare does not use the purple ink.

While studying the lobsters in Bermuda, P.M. Johnson and Charles Derby witnessed a novel approach to the use of antipredator chemicals. When attacked by a spiny lobster, the sea hare again releases two types of chemical secretions: the liquid purple ink and the gooey opaline substance. Once the lobster encounters the opaline, the spiny lobster begins a frenzied consumption of the opaline ink. Consuming the mass of opaline, the lobster acts as a child in a candy shop and devours the opaline. During this feeding frenzy, the sea hare slowly crawls away to freedom and to safer feeding grounds. The purple ink, although released in concert with the opaline ink, appears to have no effect on the spiny lobster. Why would the lobster settle for the meal of ink and let the tasty sea hare crawl away? The sea hare has developed the key chemical or chemicals to trigger the feeding reflex of the spiny lobster. Maybe the gooey nature of the opaline provides the lobster with the feeling of having just consumed a meal. The sea hare manipulates the lobster's own sense of smell to fool the lobster into believing that a savory meal is at hand. In this way, the sea hare survives to see another day.

3.2 Advertising the Home-Cooked Meal

Mornings in the bakery always seem to bring the most intense periods of writing for me, but just as all good things must end, my initial writing frenzy begins to wane after a couple of hours. Finishing my breakfast, I pack up my computer and walk to my office. To get there, I walk about 20 minutes from the middle of town to the eastern edge of campus. Bowling Green is a small town of about 30,000 residents and, by far, the largest economic mainstay in town is the university. Being a university town, the majority of downtown businesses are bars, restaurants, coffee shops, and bakeries. I estimate that nearly half of the businesses along my walking route sell food or drinks.

Passing the numerous restaurants, I notice that each of them has a unique way of advertising their latest food or drink offerings. Given the diversity of interests and people at the university, it is most likely that each of these establishments has its own clientele base. The first place that I pass is an old fashioned, greasy-spoon diner. Outside is a 1960s classic sign saying "Hamburgers, Chili and Fries" with each of the words highlighted by a circle of blue neon. Open late night, this is a favorite spot for post-party fare.

Next, I pass by one of the more recent additions to the center of town: a Mexican restaurant with an ad for "half-priced margaritas during happy hour." Right next to the Mexican restaurant is a small coffee shop with a sign describing their latest soup creations and different types of coffee roasts. Further down the road is a restaurant from a national chain claiming "The best buffalo wings in town." All of these restaurants are using carefully chosen words to evoke vivid mental images of delicious food, exquisite drinks, and rich desserts in order to draw the potential customer to their particular establishment. Some of the more modern signs also include mouth-watering pictures of the treats that can be found beyond the doors that beckon.

Just like the restaurants in my hometown, nature has its own dining establishments. In a fashion similar to the restaurants' financial dependence upon drawing in many customers, the restaurateurs of the natural world (i.e., flowers) must also entice potential diners to sample their offerings. In the natural world, there are no billboards, neon signs, or flashy words (as we know them) in which to market a potential meal to hungry animals. These restaurants that I am referring to are the world's flowers, and the potential patrons are the host of organisms that visit flowers to obtain nectar and other valuable resources. Instead of using a written language or neon sign, they advertise their offerings just as effectively using the language of smell.

Flowers provide nutritious meals for nature's diners in a couple of ways. During spring and summer, they produce sweet nectar for the many insects, bats, birds, and small mammals that inhabit the flower's neighborhood. Nectar is pure energy in the natural world and is highly coveted. During the fall, those same flowers that provided the sweet nectar now offer nature's dessert, fruit, before the harsh winter sets in. There are no free lunches in our world or in the world of nature. Just as I had to pay for my morning muffin and tea, nature's patrons must also pay the flowers for providing nutritious meals. As the meal takes two different forms, nectar and fruit, so too does the payment.

During the springtime, flowers produce nectar not just to feed others but also as a way to aid in their own reproduction. Although they are sexually reproducing organisms, being stationary means flowers need help in getting the male and female reproductive parts together. This process is called pollination and occurs when the male gamete, the pollen, is delivered to the female reproductive organ, the ovary. To reach the ovary and accomplish the goals of sexual reproduction, the pollen hitches a ride on the legs of insects, the backs of mice, or feathers of birds as they visit the flowers to dine on the nectar. These animals, called pollinators, then travel from flower to flower depositing the male gametes near the flower's ovaries. As they visit multiple flowers, the pollinators involuntarily pick up and deposit male gametes on the various flowers along their route.

In the fall, those flowers that were successfully pollinated in spring produce fruits, such as apples, pears, and cherries. The fruits are really just external houses for the seeds of the plant. During the fall, the fruit ripens, falls from the plant, and, once the fleshy part decays, leaves behind a seed that will hopefully mature into another plant. Landing and growing right next to your mother is not necessarily a good place to be if you are a seedling plant. Your mother may shade you from sunlight and stunt your growth, the roots of the parent plant may inhibit the seedling's own root growth, as well as other potential problems that accompany growing next to a fully mature plant. The seed is now faced with the same problem as the immobile male gamete described above. How do the fruits and the enclosed seeds get dispersed far enough away from the parent plant? Here again, the plant uses the services of another suite of animals. In the case of small berries, birds will often consume the berry whole, which leaves the seed undigested and traveling with the bird. Once the seed passes through the digestive tract of the bird, the seed is excreted and deposited some distance away from the parent plant. Other fruits are consumed, moved, and deposited in similar fashion by small mammals, such as mice, voles, and bats. This results in their seeds being dispersed away from the parent plant. Nature's pollinators serve to help plants reproduce and the fruit consumers help by dispersing the seeds.

Probably because of our interest in fruit and flowers, this story of pollinators and dispersers is one of the best-studied and well-understood aspects of biology. For years, amateurs and professional biologists have watched bees, moths, and bats visit and revisit flowers and have marveled at the specialized characteristics that both the flowers and pollinators have adapted. Although floral odors are long known to play a role in the flower/pollinator relationship, only recently have the specific roles of smells and pollinator behavior been deciphered thanks to advances in chemical analysis techniques.

Even those of us who pay little attention to flowers recognize the tremendous diversity in flower shapes, sizes, smells, and blooms. Some flowers bloom in the spring while others wait until summer or even late fall. There is a vast range of sizes from the tiny little bluebell flowers, no bigger than the nail on your pinky finger, to the enormous Titan flower that was mentioned in Chap. 1. There are also distinct differences in fragrance. Some flowers have little noticeable smell, while others have powerful odors that can linger for days. All of these flowers have evolved with a single purpose: sexual reproduction for the plant. But why does such a vast diversity exist to serve a simple single purpose?

The answer lies in the specific relationship between flowers and their pollinators and the need for very specific communication. Just as one would not expect to walk into the United Nations General Assembly and be able to communicate to everyone in a single language, the various animals that act as pollinators also require a range of languages that each of them can understand. Thus, we find that some flowers speak a general odor language and target as many different pollinators as possible while others speak a very specific odor language that is aimed at a single species of pollinators. The wealth of information about this story is large, and a full explanation is beyond the scope of this book. (I suggest that those interested in a deeper explanation look at some of the references at the end of the book.) However, I can select a couple of the more interesting stories to demonstrate that flowers do have a language of smell and that they are capable of multilingual communication to different animals. The flowers, like the restaurants of my home town, are targeting their advertising to a very select clientele.

Pollinators come in many different forms, many of which are insects. You are probably familiar with the honeybees and bumblebees that frequent the gardens around our homes. In addition to these insects, beetles, ants, moths, butterflies, flies, and wasps are among the more common insect pollinators. Of course, some of these groups of insects, particularly the bees, are more important to the flowers than other groups. Among the noninsect pollinators, there are bats, lizards, hummingbirds, and even some small terrestrial mammals like mice. With such a diverse group of pollinators, what kind of scents have the flowers evolved to communicate that a meal is at hand?

One method is for the flower to develop a large array of smells where each smell targets a specific group of pollinators, signaling that the flower is ready and open for business. A single flower could produce the entire array of fragrances and broadcast those fragrances to any pollinator within smelling distance. Each different fragrance would be analogous to an electronic “open for nectar” sign that flashes between English for the bees, Spanish for the moths, French for the beetles, German for the butterflies, Japanese for the bats, and a multitude of other languages. We know quite a bit about these molecules that signal “open.” One of the leaders in this research area is Dr. Rob Raguso at Cornell University. Well aware of the abundant work on bees and flowers, Dr. Raguso looked elsewhere for a suitable model to apply his unique combination of chemistry and behavior. He found that model in moths. Unlike bees, which fly almost exclusively during the day, moths fly at night as well. What makes working with moths particularly interesting is the absence of the beautiful floral colors during the nighttime flights of moths. Without the ability to see colorful flowers at night, moths and other nighttime pollinators must rely on the presence of chemical cues to guide their search to ready and waiting flowers.

Many of the flowers that are pollinated by night moths generate a compound called linalool. Linalool is not a very complex or complicated molecule and is the signature smell associated with white, night-blooming flowers that are almost exclusively pollinated by moths. Included in this group are long-spurred orchids, wild gingers, moon flowers, evening primroses, jasmine, and a few others. Linalool is the “neon light” that serves to call the moths to the nectar. The powerful fragrance is lifted by night winds and is dispersed to the moths. Upon smelling the fragrance, the moths begin a characteristic upwind search pattern that eventually helps them locate the flower. Dr. Raguso has focused his studies on the hawk moth. This is a rather large-bodied moth and has been described as the “apache helicopters” of the moth world. Being a large moth means that there is a constant need to refuel and, as a consequence, the moth makes multiple flower stops throughout the night. Without the linalool odor, the flowers are ignored by the moths. Thus, this “open for business” sign is essential to ensure these flowers are visited and pollinated by moths.

Surprisingly, linalool appears to be a very common smell among flowers. This common chemical appears in a number of daytime blooming flowers that are pollinated by bees, butterflies, beetles, and hummingbirds. There are also bat pollinated flowers that produce linalool. Although common among floral fragrances, linalool appears to play a minor role in bee pollination and no role at all in bat pollination. Instead of using a multilingual sign communicating to the whole host of pollinators, these night blooming flowers are attuned to the “language” of the hawk moths.

This is similar to the old 1960s style American diner I passed on my way into my office (Fig. 3.3). The classic neon sign that claims “Hamburgers, Chili, and Fries” is bound to deter vegetarians and attract grease lovers. Certainly, the more refined palates in town may have more money to spend, but these advertising words are not aimed at them. Hamburgers and fries are targeted at a specific group of consumers: hungry students. This restaurant cannot supply everything for everyone, so they have tailored their wording and advertising to a specific group. This is similar to the case of the night-blooming flowers, who likewise have tailored their advertising (the flower’s fragrance) to their most likely patrons, the nighttime pollinators.

Taking a different approach to communication with pollinators is the group of flowers known as Angelica. These herbs are often prized for their stems and roots.

Fig. 3.3 Old diner with neon sign

The stems are used for making wonderful candies, and even the leaves impart odors to foods if used in the right way. The roots are used in medical treatments. Turning back to pollination and fragrances, Angelica prefers not to rely on a single group of pollinators. Instead of focusing on moths and developing a unique odor aimed at moths, these flowers have a complex group of chemicals that form a tantalizing bouquet for a number of different insects—a blend that reminds humans of something ranging from musk to a sweet odor. These flowers are the "all you can eat buffets" of the plant world. One of the appealing aspects of buffets is that there is a style of food for every palette: all different types of meats, vegetables, fruits, and desserts to satisfy even the finicky eaters. Angelica has a little bit of smell for every pollinator out there.

3.3 The Smell of Fear

Nearing the end of my walk to the office, I approach one of the single most defining features of Bowling Green. A north–south running railroad splits the town almost in half: both symbolically and culturally dividing the town into the eastern University section with its dormitories and student apartment buildings and the western residential and business section with Main Street and more permanent housing.

Typical of many American railroad crossings, there are numerous warning signals; the most prominent of which is the flashing red lights that signal an approaching train. This is quickly followed by a loud bell and the black and white crossing bars that prohibit cars from crossing the tracks. All of these signals are designed and placed here for the sole purpose of providing adequate warning to those that approach this crossing. Without really thinking about me personally, the people (or company) that placed these signs are essentially shouting directly to me "Watch out for the train" or "Be wary of trouble." Although the warning signals are not directed at any specific person, they are just as effective nonetheless.

Nature has its own vibrant warning signals, just as common and necessary as they are in our world. One of the most dangerous situations that virtually all animals face at some point in their life is predation. The outcome of this situation is simple and intense. Avoid predation and live to see another day; fail to avoid it, and you and your genes are forever lost.

Some animals have developed fairly sophisticated means of communicating to each other the presence of a potentially lethal predator, largely by using auditory signals. For example, some species of primates have different warning calls depending on whether the threat is terrestrial (as in a cheetah), arboreal (as in a python), or aerial (as in a large raptor). Each of these calls has separate pieces of information (look up or look down) and results in predator-specific behavior (climb up or crouch down). Crayfish have also developed a predatory warning system, but instead of using vocal calls, they use chemical signals. In the scientific literature, these signals have been called alarm signals, which in the crayfish world take two different forms. Before we get into these signals, a little background on crayfish biology is necessary. Crayfish, and most crustaceans in general, have the remarkable ability to regrow body parts, legs, claws, and other appendages. The ability to drop a claw has evolved into an effective anti-predatory behavior. When attacked by a predator, a crayfish will voluntarily drop one of its claws. The predator will smell the claw and begin to consume the claw while the crayfish scuttles away.

This process, called excising, is a fairly expensive behavior in terms of the physiological costs. Claws are not cheap and the crayfish may take up to a year or 2 to regrow a new one. Given this extended time without a weapon, excising a claw is really a last ditch effort to escape predation. When a claw is excised, internal body fluids are released into the environment, and if the claw is eaten, more tissue and chemical signals are released. Since crayfish do not drop claws unless the situation calls for an ultimate sacrifice, the presence of these body fluids, tissues, and chemical signals could potentially provide valuable information to other crayfish in the area. Basically, if a crayfish were to smell these body fluids, the chemical cues in those alarm signals would almost certainly inform other crayfish that at the very least a claw has been excised somewhere but, more drastically, maybe a whole crayfish is being eaten by a predator. Here we have a very specific signal that is being released at a very specific time just as the railroad lights, and sounds only occur during the approach of a train.

Dr. Brian Hazlett at the University of Michigan has worked on this crayfish alarm system for years. He has shown that crayfish have developed the excised claw, not only as an effective anti-predatory behavior but also as an early warning system for a potential predation event. Other crayfish, in the presence of this alarm signal, will respond with the appropriate anti-predatory behaviors. If they are near a shelter, they will quickly run and hide in the safety of their home. If out and away from the shelter, the crayfish will freeze presumably hoping to go unnoticed by a passing predator. This is a very effective system of communicating immediate danger but the production of the signals is very costly, especially from the point of view of the animal that is losing the claw. If this alarm system were to be considered a form of language, then losing a claw in order to send a message limits your vocabulary to two sentences: the right claw and the left claw. At first glance, this system appears to be rather limited in the flexibility and long-term use. There is more to this story of alarm signals and crayfish.

Crustaceans, including crayfish, have another unique feature in addition to the ability to regrow limbs. They have a bladder in which they store "urine." (I have used quotes around the word urine to differentiate between what we think of as urine [waster products, ammonia, etc.] and that liquid substance that is stored in the crustacean's bladder. More appropriately, this chemical mixture should be called bladder water or some other bland term.) While a common feature among terrestrial animals, most aquatic animals do not have a bladder, as they do not need a highly developed kidney/bladder system to store and excrete urine. Most have a fairly porous skin or gills which allow them to slowly release metabolic waste products across these porous areas, as opposed to terrestrial animals with their water tight skin which have evolved other means of ridding their bodies of waste products. Crayfish excrete some of their waste products across their gills, so why would they need a bladder? They also seem to have a number of glands situated around the bladder which may release specific chemical products into the urine.

Work on lobsters and crayfish has shown that they "urinate" only when they are in a social situation as in fighting another lobster or crayfish (see Chap. 6), or when they are scared. Through the use of a simple catheter system, collecting urine while the crayfish or lobster performs different types of behaviors is possible, and thus, context specific urine can lead to insight into how these crustaceans use these chemical signals in different situations. If you place a crayfish in a tank with a large mouth bass, and the two animals are physically separated from each other by a transparent and porous screen, the crayfish will release 20 times the "urine" in 10 minutes that they do when sitting quietly alone in the tank for 2 days.

Maybe this urine release serves another purpose for a crayfish under attack. There is an old myth that when animals are frightened, they will often either urinate or defecate before fleeing from danger. The idea is that this sudden release of bodily fluids serves two purposes: one is to lighten the fleeing animal so that they extradite themselves from the dangerous situation more quickly and the other is to leave behind a chemical signal to crayfish in the vicinity that something dangerous is near. Testing the second possibility (that the crayfish are actively signaling to fellow crayfish) is incredibly difficult, although evacuating the bladder before fleeing has been well documented in other animals. The most common example of relieving the body of weight before (literally) taking flight occurs with birds.

The "urine" collected during the bass attacks is appropriately called "stress urine," and some very interesting things occur if this "urine" is released into tanks with other crayfish. If different crayfish smell this stress urine, they immediately begin to move away from the odor source. The closer they are to the source, the faster they move in the opposite direction. They have a heightened sense of awareness and they often get their claws ready to defend themselves. The urine sends warning signals. "Run away, danger is here," just as surely as the red flashing lights at the railroad crossing warn of potential danger.

3.4 A Bitter Pill to Swallow

Down the street from my bakery is a new bar that has a Celtic theme. The inside has two beautifully designed bars in separate ends of the building. In the side of the building that is more of a restaurant, the bar is topped with a large piece of copper that has been purposefully etched with acid droppings to give the appearance a distinctive look. All around the area, the walls have horizontally placed stones that give an earthen feel to the room. At one end of the bar is an inviting fireplace where a good meal and an excellent lager can be enjoyed.

The other bar is quite different. This room is designed to be a classic sit and chat style bar. At one end of the room, there is a small stage for local acoustical groups to play to the small intimate crowds. The centerpiece of the bar is a rather large mirror that reflects the patrons moods back to them. The bar itself and the shelves surrounding the mirror are all made from a lightly stained wood. On one side of the mirror is a long list of Irish and Scottish whiskeys and single malts and on the other side a list of the beers on tap for this week. The owner does an excellent job of rotating the beers from week to week to provide a bit of variety for those daring enough to explore something other than traditional beers.

I look forward to a late afternoon hour pint or an evening sit for a couple of reasons. The Celtic atmosphere probably draws on my Scottish heritage and makes the ancestral blood in me feel at home. This is also a nice quiet bar where I can sit and chat with my graduate students without having to speak loudly over the usual din of a college town bar. Finally, I enjoy a good beer from time to time, and the diverse choices let me explore a different world of chemical signals every week. Since I enjoy brewing my own beer at home, exploring lagers and ales from professionals occasionally gives me an idea for new home creations.

The Reinheitsgebot (the Bavarian beer purity law) originally stated that beer could contain only three ingredients: water, barley, and hops. Yeast and the mechanics of fermentation were unknown when the law was put into place in 1487. The barley provides the sugars that allow yeast to produce the alcohol of the beer and the hops give the beer the characteristic bitter taste that tends to divide people into hopheads (those that love highly hopped beer which tend to be quite strongly bitter) and those that prefer smoother tasting brews. For those that are interested, beers can be quickly and effectively categorized on three different scales. The beer color can be measured on a European Brewery Convention (EBC) scale that varies from pale lagers and witbiers that are yellow in color (EBC=4) to the deep rich black color of stouts (EBC=79). The colors result from the roasting of the barley and other grains used to produce the malt. The second scale measures the alcoholic content of the beer and really measures the density of the beer. One of the scales used to measure the density and alcohol content of the beer is called the specific gravity. Since the alcohol in beer (ethanol) is less dense than water, beers with higher specific gravities have higher alcohol content. Finally, and most important for this book, is the measurement of the amount of bitterness imparted to the beer from the use of hops. The scale most commonly used is called the International Bittering Units (IBU) and can range from 0 to 100.

In Chap. 2, I discussed one of the problems with the scientific investigation of chemical signals in both animals and humans and that was the measurement of the exact nature of the chemical stimulus. I used terms like earthy, fishy, or woody to describe the quality of odors, and this same problem arises when we think about quantifying bitterness units in beer. Measuring the color or alcoholic content of beer can be quite precise when using a spectrophotometer (an instrument to measure color) and a simple weight and scale to measure the density of the beer. Bitterness of a beer is dependent upon the type and amount of hops used as well as the amount of malt (fermentable sugars) used in the recipe. Even though the IBU scale runs from 0 to 100, not all IBU 40 beers evoke the same quality for the same person. Added to this lack of precision, different people are more or less sensitive to the bitter taste. Some people are referred to as "supertasters" and have a heightened sensitivity to the bitter sense. (More on the genetics of this interesting taste phenomenon below). Still, despite the wide variation in the perception of bitterness, there is a scale for beers that contain very smooth or low hopped beers near zero and India Pale Ales (a very highly hopped beer style) near 100. I am not a supertaster, but the highly hopped beers aren't an enjoyable experience for me. So, I am not a hophead and actually prefer smoother, high gravity beers.

From an evolutionary point of view, our and other mammal's ability to perceive bitter compounds is linked to the need to parse out food that would provide nutritious and good meals from those potential food items that contain harmful or poisonous compounds. The bitter taste stands as a sentry point for the ingestion of potentially dangerous and deadly compounds. Our history is rife with examples of illicit use of these chemicals to change the course of history. Perhaps the most famous was Socrates being sentenced to death by drinking hemlock. Hemlock contains the bitter alkaloid coniine that causes respiratory failure in both animals and humans. Some plants are so dangerous that whole generations of myths and legends have arisen from potential poisoning: Wolfsbane, Oleander, and *Cerbera odollam* which is aptly named the suicide tree. A 2004 study by a French team of forensic toxicologists found that more suicides are caused by the ingestion of this plant than any other method. A large number of these poisons are alkaloid in nature and unlike some of the other chemicals mentioned in Chap. 2, alkaloids have a large range of structural diversity making these compounds hard to classify like esters or alcohols. Despite the large diversity of compounds included in the alkaloid group, almost all of the compounds produce a bitter taste.

Plants produce these harmful and bitter compounds mostly as a protection against grazing herbivores. Termed secondary metabolites (and allelochemicals) because the alkaloids do not serve a primary role in the growth and reproduction of the plant, these compounds are critical in the plants' survival against the army of herbivores that would like to consume all of the plants' leaves. The alkaloid compounds that can be produced continuously and sequestered in the edible parts of the plant or some plants (like the oak trees in Chap. 1) can specifically upregulate their production when they are attacked by herbivores. The continuously produced chemicals are called a constitutive defense because the chemicals constitute a basic element in the edible material of the plant.

Other chemical defenses are termed "induced" because a specific act of herbivory sends the plant into action where the biochemical machinery switches on those pathways that produce the toxic defense. This is akin to a biochemical version of a

Fig. 3.4 Godzilla caterpillar

siren horn used to call troops to battle, although here the troops are individual alkaloids called to the front line attack by a caterpillar. This active analogy is specifically chosen because plants are not necessarily a passive organism in this game of survival. Some plants are actually quite clever about their chemical defenses, if one can attach the word clever to plants (Fig. 3.4).

The common flower Geranium has over 400 species of flowers and is found among the temperate climates across the globe. A small and pretty little flower that has quite a dark side to its protection against herbivory. The Japanese beetle is an invasive and voracious herbivore that is relentlessly attacking gardens where the beetle has been introduced. The flowers of the Geranium produce an anacardic acid that is located within the petals of the flowers. Upon consumption of the petals, the Japanese beetle flips over to its back and becomes paralyzed. The legs and the antennae of the beetle will twitch slightly during the time that the beetle is on its back. This simple act of paralysis is enough to stop the beetle from consuming more of the plant and seems an adequate defense. The particularly dark side of this story comes after the paralysis. If left alone, the beetle will fully recover in 24 hours and go about its way consuming more plants. Thus, the paralyzing chemical doesn't actually kill the beetle outright. What does happen during the paralysis stage is that predators of the beetle are attracted to the easy prey. Additionally, the twitching appendages serve as an attraction to birds such as starlings. The Geranium uses the beetles' own predators against them by providing the predators with an easy to consume meal of paralyzed beetle. The chemical used to paralyze the beetle does nothing to the starling or other predators.

3.5 It's an Acquired Taste

Summer field work at a biological station is a real treat for me. One of the most pleasurable aspects at my main field station is the communal meals. Cafeteria style meals eaten at a leisurely pace lead to intellectually stimulating conversations over the day's field course, experimental frustrations, future research, or even less scientific topics. One of the frequent topics of conversations that I have had over the last 18 years of field work is the meal at hand. Appreciation of the day's dessert (say a raspberry cheesecake brownie) or critique of the selection of spices for the main dish (beef or tofu stroganoff) leads us down paths to personal preferences. Occasionally, a simple yet challenging question is thrown out for discussion. Since we are eating, a question could take the form of "What would be your choice if you had one last meal?" Being scientists, we aren't just satisfied with an answer "Steak," "Fried chicken," "Raspberry cheesecake brownie," we demand in a gentle and intellectual way, an explanation of why that meal would be chosen.

We recently went down this path while sitting at a table with some friends, graduate students, and colleagues. My answer changes slightly day to day or year to year because I truly enjoy the sensory experience of eating. (My enjoyment of good food probably results in the plodding runs described in Chap. 1.) A good meal is a combination of a multitude of senses. Certainly, the sight of food plays a role in the total experience. We have been culturally raised to expect certain appearances of our food, and if the image presented before us doesn't match our expectations, then this dampens our appreciation of the taste and smell of the food. For example, while making lunches in preparation of heading out for a full day of field work, one of my students walked over to my table with two slices of bread covered in what appeared to be a mound of bright purple birthday cake frosting. After inquiring about her sandwich, I was informed that the mound was beet hummus and not birthday cake frosting. Unfortunately, my mind was so focused on the frosting appearance that I could not get myself to attempt a sample of the hummus. On top of the visual appearance, texture of food is important as indicated by a continual debate between the qualities of lumpy mashed potato versus smooth mashed potatoes. I know of several people, including my wife, that say that texture of the food is almost more important that the smell and taste. Their selection of the last meal may be completely based on their sense of touch.

Of course, the main sensory component for most meals, and especially if this was our last meal, is our sense of smell and taste. (Forgive me if I do not include the trigeminal sense which is most important for Buffalo wings, jalapenos, and carbonate beverages). In answering the question about the last meal, I would start to think about what odors I would want to inhale to maximize my enjoyment and what combination of sweet, bitter, salt, sour, and umami I would want to envelop my taste buds. Personally, I love cheese, so any meal that involved some level of cheese or cheese mixture would have to be the choice. So, after a long pause, I settle on lobster macaroni and cheese. For me the cheese would be a combination of a good smoky Gouda, a touch of goat cheese, followed by a very sharp cheddar. Mixed with elbow macaroni and the meat from a lobster claw broiled with butter and a hint of cayenne pepper. The combination of smoky odors, the sweet meat of the lobster contrasted with the bite of the cheddar and cayenne would be such a delight that I would be ready to move on from this world.

If any of the descriptions above seemed attractive as you read them, you might have found yourself salivating slightly as you imagined this meal or maybe your own meal. I was almost like Pavlov's dog thinking and writing about my last ideal meal. The concept of an unconditioned response, Pavlov's dog salivating at the ringing of a bell signifying dinner, should indicate the hidden or unconscious importance of odors (and tastes) for the singular act of eating. For most of us, we take these smells as a common place occurrence and can't imagine what our meals would be like without these senses. Thankfully for science, there are ways that one can begin to understand what our meals would be like without these odors.

As I wrote in Chap. 1, two simple demonstrations about the influence of chemicals during eating can be performed at home with common household items or done in schools to help kids understand the difference between smell, taste, and even trigeminal. This is the jellybean example with a cherry, raspberry, and cinnamon jellybean. To quickly review, with the nose blocked, the cherry and apple jelly bean "taste" indistinguishable because the sugars in the two beans are identical. The only sense that is activated is the taste buds and they signal to the brain that sugar is present. The cinnamon jellybean is different from the cherry and apple because the cinnamon actually activates the trigeminal sense. Anosmic people (either those with their nose plugged, or colds, or permanently anosmic through disease or trauma) can determine the difference between spicy and non-spicy foods. In this demonstration, taste alerts our brain to the presence of sugar, and the trigeminal sense sends information on the level of spiciness. The nose is not needed to discriminate cinnamon from apple or cherry jelly beans, without it, the apple and cherry are indistinguishable. If the nose is unplugged halfway through consuming the cherry jelly bean, the odors flood our receptors and the cherry is instantly recognizable. The second demonstration is quite similar, but somewhat more shocking in regard to how we enjoy our food. If a participant is blindfolded and their nose is blocked, they can be fed slices of apples, potatoes, and even onions, and the ability to distinguish these three distinct foods is almost eliminated. Given the powerful and pungent smell of sliced onions and the sweetness of apples, this demonstration is difficult to believe, but the nose (not the tongue) is the sense that allows us to enjoy a good tart apple or the sharp flavor of raw onion. This difference in taste and olfaction is why meals and food don't seem quite as delightful when we have colds that inhibit our ability to smell.

Instead of having a cold block their sense of smell, some people are permanently anosmic. Anosmia can happen through a number of possible mechanisms. Some people are born with a defective sense of smell and, thus, never will have known any of the aromas associated with foods. Other people lose their sense of smell by head traumas or through severe allergy issues. The olfactory receptors in the nose could vanish through cell death, or trauma to the brain may damage the olfactory nerve that leads to the inner workings of the brain. There are a host of diseases where one of the possible consequences is either a reduced or eliminated sense of smell. Despite all of the surveys that indicate that the sense of smell is the first sense that we would prefer to lose if we had a choice, for those that lose this sense, the world becomes a dull but dangerous place. Almost akin to the reverse of the famous black and white to color transition that Dorothy Gale makes when landing in Oz during the 1939 classic movie. Each meal becomes an adventure in the black and white world of texture and temperature rather than the sensuous colors of taste and aromas. The dark and deep rich aroma of the morning coffee has no effect on your state of awareness. The morning bacon is loudly sizzling in the skillet, but the only thing that can be sensed is the sight and sound. A bite into a pumpkin muffin from the bakery would provide me no symphony of flavors; just a solo note of mush as I would dejectedly chew my breakfast.

The hidden power of chemicals turns our consumption of food from a necessary task into an endless choice of sensory adventures. Eating is, quite obviously, essential to our well-being. However, our choices of how to cook food and what to eat is decided more by our sense of smell than by logical (and healthy) choices on what our bodies need. A mere sniff of the plumes of food odors reveals to the chef whether more cayenne pepper is needed for the jambalaya. A taste of the soup is used to apprise the need for more oregano or garlic to enhance the stew. For the patron of the meal, a wave of the hand over the dish delivers the meal's concoction of flavors and starts the emotional reaction to the potential first bite. For the liquid refreshments, we talk about the bouquet of a fine wine or the nose of a single malt scotch as a measure of their quality. These odors are the first, and in some cases, the most powerful engagement we have with the liquor. I would hate to imagine attempting to taste a good single malt scotch without a sense of smell.

Although these examples above might seem somewhat trivial compared to navigating our modern world without our eyes or ears, the matter is not trivial for those that lose their sense of smell. Severe depression and a state of anxiety are often one of the consequences for anosmic individuals. Some medical studies estimate that up to 50 % of individuals suffering from anosmia also suffer from depression, and many of those individuals go through significant weight loss as a result of a lack of appeal in most foods.

At the other end of a theoretical spectrum from anosmia are the supertasters. Given the discussion above about the depressive nature of the loss of smell, super tasting might be thought of as a true joy, but this would not be so. Supertasters are like having taste buds on steroids. These individuals have enhanced experiences while tasting their food and drinks. The underlying mechanism is related to the presence of a single gene (the TAS2R38 gene) that is absent in regular tasters. Although all of the impacts of this gene are not quite known, the supertaster sensation is due to, in part, an increase in the number of fungiform papillae which contain taste buds. Estimates on the total number of human supertasters are around 25 %, and a simple test can be performed to determine a supertaster. Dr. Linda Bartoshuk has been at the forefront of this phenomenon in humans since the start of the 1990s. The compound PTC (Phenylthiocarbamide) or PROP (propylthiouracil) is placed onto a small piece of dissolvable paper and subsequently placed on the tongue of the testee. If this person is a supertaster, the reaction is immediate and obvious. These compounds will taste exceptionally bitter and will produce a reaction of disgust once the PTC square is placed on the tongue.

Although this group of supertasters is found by a bitter taste test, the enhanced sensory experience of the tongue has implications beyond just this one of the five tastes. Through Dr. Bartoshuk's lifelong work on the psychology and behavioral consequences of taste, supertasters have been found to have different preferences for those foods that contain bitter compounds. One of the most hated vegetables in the world, the Brussel Sprout, may be hated because of the large contingent of supertasters that are disgusted by the bitterness of these green vegetables. Supertasters will also avoid other green vegetables, grapefruit juice, and soy beans. They may have a more intense burn from alcohol and the chemical compound most commonly found in spicy foods (capsaicin). On the end of the taste "spectrum," these same individuals have an enhanced sensation to sweets. Cakes, ice cream, and candy have an enhanced flavor to them and for many supertasters, these treats are too sweet. Supertasters are perpetually stuck in food equivalent of the Goldilocks' fable. Instead of the porridge being too hot or too cold, foods and drinks are often too bitter and too sweet. They are constantly on the hunt for baby bear's food that is just right.

3.6 And Finally, Monsieur, a Wafer Thin Mint

Throughout this chapter, I have attempted to explain a wide variety of examples that show how an animal's behavior is influenced and controlled by the odorous world around them. From crayfish detecting the fear of predation from other's urine to flowers tuning their chemical advertising to the specific wants and needs of their pollinating customers, animals and plants have evolved elaborate uses for chemical signals in their quest for food. In our own world, the smell of a marshmallow toasted just right over a roaring fire or the odor of grilled onions placed on top of a broiled streak triggers both overt behaviors (grabbing a graham cracker and chocolate) and hidden physiological responses (the enzymes in our digestive system). Given the wide range of food related odors that influence our lives, I can leave you one last tasty morsel of a chemical story or as Monty Python would say, "a wafer thin mint."

One of the most common and powerful odors that influences our physiology and mental state is that of mint. Derived from plants in the Lamiaceae family, mint essential oil and menthol are used as flavors and fragrances throughout the human household. Mint is found as flavoring for candy, but also can be used for medicinal purposes (treating stomachaches) and as natural insecticides. Peppermint has also been used in hospitals to reduce some of the nausea that occurs after coming out of anesthesia for a surgery. These uses are fairly well known and are not that surprising. Yet, the impact of mint (and its odors) upon both our psychology and physiology is a great example of the hidden power of the chemical senses.

Peppermint, either the flavor or the scent, has the ability to improve mental acuity when performing tedious tasks. Dr. Bryan Raudenbush and his students have performed an interesting series of studies demonstrating that different cognitive abilities are enhanced following exposure to peppermint odors. When challenged with computer tasks that required counting, puzzle solving, and memory skills, those participants that inhaled peppermint improved on their ability to focus and performed these tasks much more quickly. In similar studies, the odor of peppermint decreased the perceived workload or frustration when participants ran on treadmills. These studies show that food odors (even at very small concentrations) can control our moods and focus. Far beyond just the necessity of nutrition and health, the smells and tastes of foods are entryways for odors to influence everyday tasks and everyday moods.

Chapter 4
Who Are You?

Recognition of Self and Individuals

My office is almost exactly equal distance from either end of the biology building hallway. Coming in from the parking lot, I enter the building and pass by the department's main office. As I glance in the open doorway, I can easily recognize the faces of my colleagues. Since there appears to be an early morning gathering, I walk in to greet each of them by name. One of my friends is about 6 inches taller than myself and has brown hair that defies any normalcy of combing. He has a self-described "Bostonian" nose that fits well with his longish face. Seated next to him is another of my close friends. All of his facial features, if taken individually, are well within what we would term average. His nose is neither large nor small. His eyes are neither round nor almond shaped. Yet, even though each of his features is wholly unremarkable, the composite picture is unique, well-defined, and unmistakable.

I pause for a second and begin to scrutinize each of the faces present in the office. For just a second, I focus my attention on a single but most prominent facial feature: the nose. Noses, in relation to our whole bodies, are small and yet play a role, as central as their location on our face, in our characterization as unique individuals. In this office alone is a suite of diverse shapes and sizes. There is the large bulbous nose that adds a touch of W.C. Fields or Winston Churchill to one person's face. Another is long and thin and, for some strange reason, reminds me of a stereotypical British school headmaster portrayed in many films. A third has a concave shape, the classic ski slope. Some noses are large and round while others are mere bumps. My own nose has a slight ridge in the middle, a remnant from an errant grounder on a bumpy softball field.

Although the nose is a prominent feature on the face, its functional use, smelling odors, appears to play a very small role in the recognition of others. I cannot recall any characteristic odor of any one of my friends present within the office. If we were suddenly plunged into a blackout, I would fail miserably at identifying these individuals. This is one of those areas that we have failed to use any of the olfactory abilities that nature is using so prominently. Still, I sniff to catch any faint traces of telltale perfumes or colognes that can be attached to someone present. My initial attempt fails to produce any odors, so I take a deeper smell while trying not to be too

P.A. Moore, *The Hidden Power of Smell*, DOI 10.1007/978-3-319-15651-4_4

obvious. With this sample, I can detect a very faint trace of a cologne, but I have no clue to whom this odor belongs.

By broadening my gaze beyond their noses, I see an even greater breadth of characteristics that make us individuals. Blonde, brown, or red hair, long or short hair, curls, spikes, and bald palates all contribute to distinction. When a good friend of mine from Australia came over to the United States to start a new research position, he had long, black, slightly curly hair that reached the middle of his back. This was his trademark. After being here a year, he opted for a change in his life and promptly buzzed his hair to a length of ¼ of an inch. At first, I had to pause for a second or two to recognize him without the signature ponytail. Yet, over time, my pauses became shorter and shorter and eventually, the new shorter haired image replaced the one with the ponytail.

Other physical features can also be used to identify individuals, height, eye color, body shape. In addition to these obvious visual signals, more subtle differences in voice allow us to recognize individuals even when we cannot see them. My office is further down the hall from the main departmental office and when colleagues and students pass by with hellos, I can easily imagine their identities simply upon hearing their voices. Tone, volume, pitch, and even the tempo of the single word "Hello" can provide me with enough information to distinguish my friends. The departmental chair that hired me has a very distinctive voice, so unique that I find the description hard to construct: part gravel, a hint of high tone, and a weird mixture of softness and firmness. After talking to him a single time on the phone, I could recognize his voice the second time he called me. One of my colleagues sounds as if he has a resonator within his chest. I can only imagine how he sounds lecturing away in one of our bigger classrooms. You can hear him laugh all the way down the hall. Our minds are so in tune with the subtle differences in voice and style that we can often recall those voices just by mentioning their names. Think of all those distinctive celebrity voices, such as James Earl Jones or Fran Drescher. Our ability to recognize sounds and visualize the image that accompanies them allows us to instantly recall their faces even when we just hear a few phrases. As further evidence of our ability to recognize individuals, celebrities such as Rich Little and Dana Carvey have made whole careers out of "fooling" our ears into believing we are hearing someone else.

This is how we recognize those individuals that play an important role in our lives. The deep-bodied voice, the light golden hair, or the ski slope nose are familiar and comforting indicators. Sights and sounds carry to our waiting eyes and ears the subtle differences in friends and family. Often we take this ability for granted; imagine what life would be like without the ability to recognize each other. (One needs only to talk with an Alzheimer's patient or caregiver to realize how quickly these abilities can be lost and how painful the consequences.)

Using our perceptual abilities, we may even be able to recognize some level of relatedness in strangers. How often have you sat in a restaurant and noticed people that "must be sisters/brothers"? Or seen a child which clearly has its mother's nose. Even though each of us has a unique combination of genetic makeup and environmental influences, supplied by our mother and father, there are always strong traits that show through. The distinctive ears of the British royal family and the recognizable face and New England accent of the Kennedy family are examples of such relatedness. As in previous chapters, recognition goes beyond the sights and sounds of our friends. We shall see how the different organisms in nature have capabilities that far outstrip our meager abilities to recognize individuals using only the sights and sounds of our world.

4.1 Who Am I?

Several different types of recognition can be found in nature, some obvious and others not. Regardless of the specific type, they can all be based on some level of biological organization. Some of the levels of organization in nature and their subsequent types of recognition may seem quite trivial to us, but we need to step away from visual biases toward the needs of nature. In order to delve into the many roles that chemical signals play in the ability of organisms to recognize others, a system of the different types of recognition is necessary. To fully appreciate the roles of recognition, a more nuanced view of the inherent complexity of the biological world is needed.

Nature can be viewed in a hierarchical fashion of increasing in size and complexity. This view of nature allows a level of organization that is helpful for understanding different levels of recognition. The smallest of the biological units, important in the context of this book, are cells: the basic building blocks of life. Cells define life, namely who we are and what abilities we have. To say that without cells there would be no life is not too extreme; however, all cells are not created equally. There are large-scale differences in cellular structure, organization, and function. Some cells, such as those in unicellular organisms, perform every function that is necessary for life. Other cells, like some of those in our highly specialized bodies, are very adept at a single function but inept at all others. Thus, at a very basic level, organisms (and the cells within them) require the ability to recognize one type of cell from another. In a broader sense, this can be thought of as recognition of self from nonself.

At first glance, the concept of recognizing who you are and who you are not seems very trivial indeed. When I awake in the morning and plod down to the bathroom, almost as a force of habit, I look into the mirror. Maybe to see if I need a shave or to see if I am still alive, but I always recognize myself. Yes, that is me staring back. A little older each day and maybe a new wrinkle every so often, but without fail I know that face staring back at me is me. Holding up my hands, I see the boundaries of my fingers. I can easily recognize where I begin and end. If the bathroom mirror could provide us with an inner view of our body, we might find that the ability to recognize ourselves is not so trivial after all.

When we look at ourselves, we are really perceiving the composite image of many more than a million cells. Humans and all of the other larger organisms on this planet are multicellular; different groups of cellular units work in concert to perform the many functions of our daily lives. Brain cells are used for thinking, muscle cells for movement, epithelial cells for skin, and so on and so on. For multicellular organisms, a basic imperative is the recognition of our differentiated cells apart from each other. The functioning of these differentiated cells varies across our body and allows our organ systems to perform complex bodily functions. In order for these functions to be carried out, the right commands (i.e., release adrenaline in response to fear) need to be sent to the right cells (the adrenal glands instead of our lungs). Instead of a body composed of cells upon cells, imagine that each of us is a large city composed of millions of different people. The different cells in our body are now the different workers that are needed to run all of the systems in our city. There are workers at the power plant providing electricity to all of the other workers in the city, similar to our digestive tract providing chemical energy to our various cells. Sanitary engineers keep our streets clean by removing trash from the city, just like the cells within our kidneys and urinary tract. City hall is located in our brain with the numerous nerve cells functioning as the politicians. (Thankfully, our brain is far more efficient than a city hall.) Our circulatory system is the highway of the city full of trucks and drivers delivering goods across the city. Finally, the police and military are the white blood cells protecting our cities from outside invasion. If the city is invaded by an alien menace, the police and army jump into action and begin to repel the intrusion. But how effective could our city's forces be if the ability to differentiate the alien invaders from the happy workers is impossible? Our army would be useless without the ability to identify and recognize the difference between friend and foe, for without this ability the city would fall into ruins (Fig. 4.1).

These alien invaders are not part of some science fiction movie gone rampant but are the infamous bacteria, parasites, and potent viruses of nature. These organisms would love to usurp our biological machinery for their own nefarious ways, and, if successfully invaded, our bodies can become ill and maybe even develop a long-term disease. These invaders may cause something as innocuous as a cold but or something as deadly as AIDS or Ebola. Our body's police and army have the abilities to isolate, quarantine, or, in some cases, destroy the alien invaders, but without

Fig. 4.1 City of cells and organs

the ability to recognize an invading bacteria cell from a kidney cell or the parasite from the host, our immune system and white blood cells are powerless. The mode of detection of alien from non-alien is done through chemical signals which function as a national ID card for our bodies. In some cases, this cell/cell or self-/nonself-recognition can go awry, and this in part explains some aspects of our more serious diseases (more on this later).

As I shift my focus from this inner view, I wonder what would happen if my cells were not all housed within the shell of my body. What would become of me if my cells all had minds of their own? Along this vein of thought, there is a unicellular species that temporarily functions as a "multicellular organism." *Dictyostelium discoideum* is a very long and complicated name for such a small organism. Dicty, as it is known to it most ardent fans, is commonly called a slime mold. Being unicellular, Dicty has no muscles, no eyes, and no ears. This organism is just a single membrane that contains a number of different organelles. (Organelles are the subunits of cells and perform the basic biological functions that are necessary for cells to live. As the name implies, they are the cellular equivalent to our organs.) Dicty has such strange properties and behaviors (yes, a single cell can exhibit behavior) that this organism has become an essential element in the search for an understanding of the mechanisms involved in cell recognition, cell movement, and cellular determination (the process by which embryonic cells develop into functioning brain, muscle, or other cell type). So important is our microbe that the National Institutes of Health chose Dictyostelium as one of the first organisms to have its genome sequenced.

4.2 Come Together, Right Now, over Me

Dicty moves in amoeboid fashion, essentially a blob that slowly stretches out one section of its cell, firmly grasps the surface, and pulls itself forward. We can find slime mold just about everywhere on the forest floor, living among the leaves and ground litter on the soil. Dicty lives out its days slowly sliding along the leaves and twigs, eating bacteria, or reproducing. There are millions of slime molds in common forest litter and, thankfully for Dicty, bacteria are common in and around the leaf litter. Just as in other biological systems, there are periods in which food resources may become scarce and the good times are gone. If you were a tiny little slime mold living on the forest floor, what would you do if you were to run out of food?

We find that Dicty probably does what we would do if we were hungry and had no food: you contact your friends. During these periods of starvation, Dicty sends out its chemical equivalent of an SOS message. Waves of pulsed chemicals travel out in circles from our starving organism. The chemical (cyclic adenosine monophosphate [cAMP]) used by Dicty is ubiquitous within the world of biology and is found in virtually every nerve cell within our own bodies and is an essential element in many different cellular functions. cAMP is a critical link in the sequence of chemical events that allow our neurons to become active and to signal to each other.

Once cAMP is "smelled" by the surrounding Dicty cells, a miraculous and wonderful transformation occurs. If you were to put a Dicty cell underneath a microscope and watch its normal behavior, the cells would look like a drop of coffee spilled on the counter. Under the microscope, you would see that the cell has an amorphous shape with small blobs (arms) sticking out in various directions slowly pulling the cell this way and that way. Once the chemical call for help goes out, these cells assume their racing shapes and instead of a blob, they form a streamlined linear shape, almost like a toothpick. As if someone has yelled "CHARGE!" all of the surrounding cells surge forward, as fast as possible, toward the center of the circular waves of the chemical signal (cAMP). Once there and once 100,000 or so of their friends aggregate, this virtual city of independent cells forms a collective being and becomes a multicellular organism. A series of transformations occur that cause a group of single cells to collect and form into a mound. Here, by some unknown mechanism, cells choose to differentiate into two different types: prestalk or prespore cells. The prestalk cells are usually located on top of the mound, whereas the prespore cells are on the bottom. The consequence of this "decision" is that the genes of some cells, the prestalk cells, will live on to the next generation while others will not. At this point, previously separate and independent organisms are now communicating and coordinating their efforts into a single collective function similar to the way our own body cells communicate and coordinate. Without the ability to recognize and coordinate their efforts through the use of chemical signals, these tens of thousands of slime mold cells would die of starvation (Fig. 4.2).

At this point, the mound begins to form a skin-like surface, then slowly elongates and topples over to form the slug stage. The slug stage empowers the Dicty, giving the group of cells the mobility needed to escape the perils of famine. The slug form of Dicty exhibits a variety of behaviors, including sensitivity to light, heat, and chemicals, which are not present in the single cells. The slug, sliding on its path of slime, makes its way to the top of the nearest leaf to perform its last magical transformation. Dicty alters its cellular organization again to form a long, thin stalk.

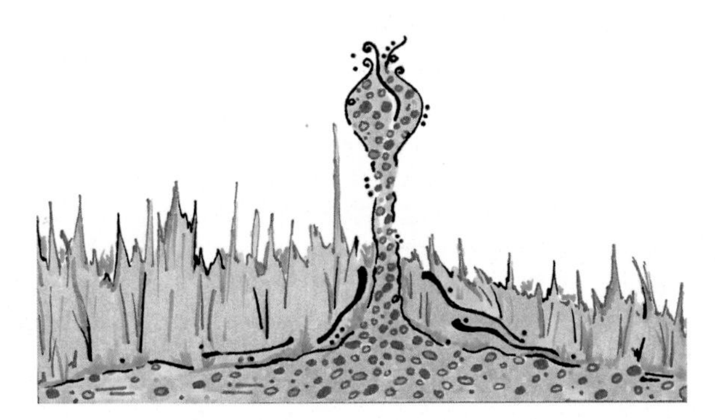

Fig. 4.2 Slime mold stalk

At the top of this stalk is a fruiting body that will release its spores to the wind in hopes of locating a more beneficial and abundant habitat or at least one with abundant food. Leaving the stalk and the rest of the slug behind, the spores alone carry the hopes of the whole "multicellular" Dicty organism for future existence. In the Dicty, we see nature's ability to solve the problem of starvation through the ingenious use of chemical signals. Through effective cell/cell communication and recognition, the *Dictyostelium discoideum* lives on for another day. But what would happen if the chemical phone lines are down?

Simply put, many physiological diseases, AIDS and cancer for example, are cases in which cellular communication malfunctions. Either our body fails to recognize the difference between our own cells and those of the invaders or individual cells no longer recognize and respond to chemical signals that mediate normal cell growth. The key to understanding and solving these diseases may lie in the ability to restore chemical communication in the body to its natural state.

Diseases are the result of cell communication gone awry, and there are even more nefarious ways organisms can intercept chemical messages and use them for purposes of which they were not intended. Parasitism is not necessarily an enchanting way of life, but from a biological perspective, parasites show remarkable versatility. At the base of every parasitic lifestyle is the effective commandeering of one organism's bodily functions (the host) to support the life of another organism (the parasite). This usually involves the parasite invading the host body and subsequently living off of the host's bodily fluids. Of course, the host does not always appreciate house guests, and its immune system often attempts to fight off the parasite and an intense physiological warfare ensues. A parasite's victory tends to lie in its ability to suppress the host's immune system or the self-/nonself-recognition mechanisms.

In World War II, the American and British forces invested an enormous amount of money and effort into both generating complex codes in which to embed hidden messages and to break the equally complex codes sent by enemy forces. Success was dependent upon the tremendous mental abilities of the code breakers, who often provided the American and British forces advance warnings of potential attacks. The code breaker's challenge was much like that faced by parasites. Parasites must understand and break the incredible chemical code of communication that allows cells within a single body to communicate, consequently allowing cells to recognize those that belong and those that do not belong. Once behind enemy lines, the parasite must break the code of chemical recognition and then begin generating false signals in order to instruct the enemy's soldiers (white blood cells and the immune system) to ignore its presence. This is a complex task, indeed, for the little parasites. Biological spies, nature's parasites, must become exceptional at cracking the chemical code of cellular communication or else risk their lives during invasions.

There are countless examples of successful parasites, but one group in particular is a master of unparalleled success. The rhizocephalans, or rootheads, are a group of barnacles that live on crabs as adults. When any other parasites attack crabs, the host crab cleans itself, a behavior that has evolved over time to remove parasites. Yet when a root head comes along, even though the crab is still capable of performing the cleaning and grooming behavior, the crab does not appear to recognize the rhizocephalan as something other than itself. The root head seems to have found the secret code—a chemical signal that either suppresses the immune recognition system of the host crab or even sends a false chemical signal to the crab that says "I am part of you." This parasite is not satisfied with simply fooling the crab with one simple chemical signal, rather the rhizocephalan is set on total domination of the host crabs body. Once the parasite has been successful at fooling the crab's perimeter chemical defenses, the parasite infiltrates the body of the host and begins to commence into phase two of the attack. The story of the root head has been worked out by Dr. Jens Hoeg at the University of Copenhagen.

Crustaceans, including crabs, are defined, in part, by the presence of a protective shell. Although an excellent adaptation for defensive reasons, the shell can get tight and cramped when the crustacean begins to grow. So, crustaceans go through a molting process where they turn in the old shell and grow a new one. Once rid of the old home, they swell up with water and "stretch" the softer new shell to provide some growing room. As the newer shell is exposed to the environment, the shell begins to harden and finally forms into the strong shell. The precise timing of the release of different hormones and chemicals controls this complex process of growth and molting. A side effect of molting is that any parasites or other animals that have attached themselves to the old shell are discarded, and for a root head, getting left behind with an empty shell just will not do.

To combat this process, the root head infiltrates the nervous system of its host and stops the molt cycle. [Thinking back to earlier in the chapter, the inner workings of a multicellular organism were compared to a large city where each worker requires precise communication between the groups of workers (or cells) to coordinate their individual efforts.] To stop the molt cycle, rhizocephalan needs to break the code of cell communication and begin to control what and when signals are sent. Since molting is controlled by precise hormonal methods, the parasite has yet another successful decoding session and starts sending false signals that would make any spy proud. (As a side point, the parasite also castrates the adult crab by commanding and controlling its nervous system, again, probably by chemical means.)

As the coup de grace of chemical communication, root heads cause male crabs to become females. This transformation is complete in both the appearance and the behavior of the crab. But whatever for? Many female crustaceans carry egg sacks on their abdomen which they try to protect and nourish through certain behaviors, including fanning themselves to help oxygenate the eggs and grooming themselves to remove parasites, fungi, and other egg pathogens. While most of the root head is inside the crab and functions to chemically control the animal, the parasite produces eggs and locates them on the abdomen of the infected crab, precisely where the crab's own egg mass would be. The root head manipulates the crab's own reproductive behavior, which evolved to benefit the crab's eggs, in such a way as to now benefit the health and well-being of the parasite larvae. The crab grooms and aerates the egg mass as if the mass was its own. Again, because this behavior is generated by the neural and hormonal signals in the crab's body, the root head has decoded these hormonal signals, seized control of the communication pathways, and has begun to generate chemical signals that control the crab's behavior for the benefit of the parasite. Finally, as the larvae mature and are ready to hatch, the rhizocephalan commands the crab to perform typical crab behaviors that help release and liberate the larvae from the underbelly of the crab. The larvae are scattered to the ocean currents to infect some other unsuspecting crab. Knowing how important the recognition of cells and self is to the existence of our lives, we can be rest assured that there are no human parasites that are as chemically sophisticated as the root head.

4.3 Who Are You?

It is 5:30 in the morning and the alarm is ringing. After tapping the snooze button one time, I roll out of bed. Almost routinely, I am the first one up in the house. Skipping the run this morning I hop into the shower and start another day. After a thorough cleansing, I return to the bedroom to get dressed. My wife is still asleep, but an unconscious morning ritual has occurred yet again. She has rolled over from her side of the bed to my side. Although she may not admit to it, I have some idea as to why this simple action has become a common morning occurrence. The secret lies in the nose.

On those mornings that my wife precedes my exodus from the bed, I, too, roll over to her side. There is nothing unique about our pillows; both are identical size and shape. She (and I) do not spread out to take advantage of the now spacious bed. We both simply roll over to the other side. Neither side of our bed is fluffier, warmer, or firmer than the other side, but there is something that the "other" side of the bed holds that my own side can never have: her individual and unique smell. Upon rolling over to her side, I embrace her pillow and am immediately wrapped within her smell. This is her. I have come to learn and love this fragrance. To me, the odor is her smile, her warmth, her laughter, and wit all wrapped into one single sensation. All carried by tiny molecules that drift from the pillow to my nose. This smell is her identity that I could identify anywhere.

A common advertising ploy is to perform blind taste tests comparing two products: Coke versus Pepsi or diet versus regular soda. We have all seen them on TV. The unsuspecting participant is handed two cups, each with a dark bubbly liquid in them. Upon tasting both, the participant declares that cup "A" is obviously the one that they commonly drink. "This is the one that I drink. The regular soda," they announce. Lo and behold, when the host reveals which is which, the participant is wrong and they have actually picked diet Coke over the regular Coke, or some variation upon this theme. Although this example involves taste, I am as equally confident that if presented with a blind pillow test I would easily be able to identify my wife's pillow using my nose. Yet, just like our imaginary participant, I would probably fail if presented only with her natural odors minus the fragrances of perfume, shampoo, and other artificial smells.

Our morning ritual is an anecdotal example of individual identification using odors. Individual recognition is the ability to detect an odor and to identify either the individual or some aspect of the individual (more on this a little later). For my wife, her odor is the distinct combination of perfume, shampoo, and deodorant mixed with her body's own fragrance, trapped upon her pillow every night during her slumber. Through our numerous close exchanges (hugs, walking together, etc.), I am exposed to this odor and, through time, have come to associate this with her. Her odor is not the only one imprinted in my mind. Every time I kiss my daughter or son on their head, I sneak a quick sniff and delight in their smell. As good as a bloodhound, I believe that I could pick their odors out from any lineup. Although they use the same shampoo and soap that we do, each of them has a unique odor unto themselves. Again, through our hugs and play sessions, I have come to learn and memorize their smell as either my daughter or son.

Our individual odors, unlike the rest of the animal kingdom, are generated through a mixture of self-produced scents, or body odor, and those artificial scents we use to either hide or enhance them. Chanel No. 5, strawberry shampoo, lavender soap, and mountain spring deodorant are just a few examples of smells that we have around the house. Each of these mixes with our own body's odor to produce the combination of smells that says, "Here I am." Nature has no perfumes, no fragrant soaps, and no store to purchase these artificial enhancements. Individuals of a species must use other means to become unique or to be identified as individuals.

Many animals have either the ability to identify individuals or maybe some aspect of the individual through odors. Just as I can identify my wife and kids through their odors, animals have the ability to recognize individuals through their odors, and these odors can carry very specific and important information. Some species have the capability to associate specific odors with specific individuals. Through association and learning, an odor can carry a message like "Hi, I am Bob and we met last week." Concurrently, a specific individual odor can carry information that allows the recognition of common trait, social position, or group affiliation that is essential for carrying on interactions with other animals. Odor can convey information on whether the animal belongs to a specific group, such as the ability to recognize nestmates, as was seen in the ants and their thieves from Chap. 1. Odor recognition can be such a very specific phenomenon that it essentially allows the animal that is smelling to penetrate through the outer layers of skin and muscle and perceive the very DNA of the animal sending the odor. These odors can express some form of relatedness, such as the ability to recognize siblings, parents, offspring, or other relatives. For those animals that live within very ritualized social systems, such as ants, primates, and even aquatic crustaceans, odors also provide information on other types of recognition, such as recognition of the sex of an individual or even the species. These may seem to be simple tasks at first glance but are essential for the survival of the individual and species.

4.4 Hello, My Name Is Inigo Montoya

The concept of individual recognition is prevalent within several large circles in biology, and a number of researchers have shown or hinted at the abilities of animals to recognize and differentiate individuals through odors alone. Given our own abilities to differentiate between people using other sensory modalities (as described in the beginning of the chapter), the thought of other animals being able to recognize

Fig. 4.3 Bosses name tag

individuals may seem ordinary. Describing the evolutionary advantage is also fairly straight forward, but individual recognition (with all of the undertones of social status and genetic relatedness) is certainly not as trivial as our own ability to recognize individuals by their facial features. Once animals have developed the ability to smell and identify individuals, a relatively small step forward is the development of very complex social systems based on individual recognition. In fact, one of the key elements in a stable social system is the ability to recognize and remember not only individuals but their status or social standing. Imagine how much trouble would be caused if we could remember the face of our boss, but forget that he is the boss (Fig. 4.3).

Many animal behaviors and traits are used to maximize survival and reproductive efforts. In evolutionary terms, these two things, survival and reproduction, are tied into the concept of evolutionary fitness (surviving and reproducing OR passing on one's genes). A quick glance at the diversity and wonder of nature shows us that animals have capitalized on a variety of methods to maximize fitness. Specifically with regard to maximizing reproductive efforts, it takes two to tango, as the saying goes, and this holds true for all sexually reproducing species. For many organisms, including some plants, bacteria, and animals, a solo tango works just as well as dancing with a partner. Even more complicated, some sections of nature have chosen to participate as both soloists and as part of a duet. Inherent in this choice of a reproductive partner and the consequences of that choice, we find the evolutionary advantage to those who can recognize individuals. Clear long term fitness benefits are provided to those animals who can recognize good mating partners (i.e., those that are most fertile and healthy) but also to those animals that can continue to recognize their mate and the offspring of the mating. Imagine an animal that chooses a partner, mates, and then leaves for a short period to forage or to chase off an intruder. Upon returning, if this animal cannot recognize their mate, all of their nest building, gathering of food, and protecting goes to waste. Worse yet, if they accidentally help a different nest, the effort goes toward another animal's evolutionary fitness. Although recognition may seem quite obvious, individual recognition allows animals to invest time and resources into those individuals that are related to them or those individuals who share a common evolutionary fitness, thus benefiting the individual either directly or indirectly.

A number of organisms are gregarious and have very complex social behaviors and social hierarchies. Although, the diversity of social systems found in nature is vast, at the heart of all these social systems is some sort of differential standing within the society. Primates have alpha males and females that obtain the preferred food, shelter, or mates, and social insects have queens that are the only ones who have the ability to reproduce. In mammalian systems, the social standing often arises, in part, from antagonistic interactions or from some level of aggression. The ability to win these interactions moves an animal up the chain of command while losing has the opposite effect. Whether the outcome is winning or losing one of these matches, the participants expend a significant amount of energy and can be hurt in the process. Within these social systems, there are numerous examples of individual recognition within a social context. If an individual has the ability to recognize a former opponent and remember the outcome of the encounter, they can possibly reduce their energy expenditures or the likelihood of receiving injuries. By being able to recognize a former opponent, a chimpanzee can say "There is Jim and I lost a fight to him yesterday. I think I will avoid him today." Another one may say, "I recall winning against Anne yesterday. I think I will go over just to reinforce my position." Recognition of individuals allows animals to make a more informed decision about whether to fight today or move on. As we shall see below, there are even more benefits to social recognition of individuals within your community. From these simple thought analyses, we can see a tremendous advantage in (or in evolutionary terms, large selection pressure for) the ability to recognize individuals.

Here we are faced with a problem that needs to be solved. How do you smell and remember something as unique as an individual? Organisms change their physical appearance through growth and development, and vocalizations often change through these same processes. Even the subtle shades of fur and skin change appearance throughout the day. During early sunrise or sunset, sunlight can often take on red hues, and cloudy days make the skylight appear gray. As we saw in Chap. 1, the catfish body odor can change through alterations in their diet. In order to smell an individual's identity in the light and in the dark, signals need to have two specific properties. First the cue or signal must be unique within the enormous world of natural diversity and, second, this cue must be constant.

There is a natural substance that fits this description perfectly: DNA. Each organism's suite of genes is unique throughout all of nature and remains essentially unchanged from birth to death. Smelling an organism's DNA would provide a perfect tool for recognizing individuals. So unique is the structure and pattern of an individual's DNA that DNA fingerprints have become a standard practice in law enforcement as an evidence of proof. Within the broader scheme of nature, studying the unique patterns of DNA has opened up new areas of research regarding evolution and speciation and has become essential for understanding the conservation of endangered species.

The only problem with detecting DNA is that this molecule is trapped within the body's cells and is not really accessible to the nose. Perhaps instead of smelling the structure of DNA, a signal could be related to an individual's DNA and maybe organisms can smell a product of DNA. Something that is as unique as DNA but that produces a signal that can also escape the boundaries of the cell and body. There just so happens to be such a thing and this system is called the MHC (major histocompatibility complex).

The MHC is a large region of the chromosomes that contain several closely related genes. The MHC is a section of DNA that works to control how our immune system performs self- or nonself-recognition. To truly understand how recognition could function, understanding how the MHC makes chemical products is necessary. The MHC produces a class of proteins called antigens that are inserted in the membrane of all of the cells in our body. This, as alluded to earlier in the cell recognition section, is how our immune system knows how to attack invaders to our system and to leave alone cells that are produced by our own body. Self-recognition is a key point to how our immune system can function so effectively and keep organisms healthy. Antigens are actually proteins and consist of sequences of smaller molecules called amino acids. Antigens can be hundreds to thousands of amino acids in length. Amino acids are the alphabet of proteins, and the structure and function of each protein is due to the types of amino acids present and the order in which they occur within the protein sequence. Change an amino acid and the protein is changed. The genes in the MHC complex are highly dense and can be found in multiple combinations. Given these properties, the MHC complex of genes solves the two key requirements for an individual odor. First, these molecules are unique and, second, the molecules remain consistent throughout development.

Insight into the role that the MHC plays in individual recognition has been provided by the work on mice by Drs. Kunio Yamazaki and Gary Beauchamp at the Monell Chemical Senses Center. Mice have the ability to discriminate between two different mice. Now, this may seem like nothing special. Of course, if the test mouse is provided with a large enough difference in the sensory perception of the two experimental mice, we would expect them to be able to differentiate between them. The key question is how sharp their senses are. Inbred mice consist of genetically identical mice. This means that mouse A has gene A then all other mice in the strain will also have that gene. Congenic strains of mice are derived from the inbred strain and are genetically identical to each other except for a single gene. What exact gene is different varies based on the congenic strain used. Dr. Yamazaki has shown that his mice are able to discriminate between different congenic strains of mice. If the mice are provided with only odors produced by congenic strains of mice, they still respond differently to the two strains. Also, the whole mouse is not needed to produce the key scent for discrimination. If you just provide the urine cues to the test mice, they perform just as well. Mice are pretty smart creatures and can be trained to perform all sorts of tasks, yet no amount of training could teach the mice to discriminate between two inbred mice. The mice are smelling the difference between the DNA of one animal and the DNA of another. How do Dr. Yamazaki's mice differ genetically? The only genetic difference in the congenic strains was located among the MHC complex. In addition, Dr. Yamazaki and Dr. Beauchamp showed that these mice have the capability to discriminate between two MHC molecules that differ only by three amino acids. Imagine reading the last three pages twice and noticing that the passages differ only by three words!

If the MHC makes proteins that are bound to cell membranes, how does the MHC influence body odors to the extent that allows for this level of individual identification? Urine appears to be the essential answer to this question. The MHC produces proteins that are not only found on cell membranes but some of the antigens are also found in the lymph, blood, and urine. The MHC proteins themselves are rather large and are not that volatile. One of the key properties of a potential chemical signal molecule for terrestrial animals is that the signal can be transported through the air from the sender to the receiver. Large nonvolatile molecules are just as useless as a perfume that doesn't travel through the air. These large MHC proteins are associated with some smaller molecules that are carried into the urine along with the MHC antigens. Once in the blood stream or in the urine, physiological processes begin to breakdown these molecules into smaller and smaller unique chemicals that are very volatile. The genetically unique mixture of MHC antigens would carry a unique mixture of smaller molecules into the blood and urine, thus providing an individual odor associated with urine. Although we tend to think of urine as something only worthy of flushing down a toilet, the animal world has developed urine as the single best book for the language of smell. As we shall see in future chapters, urine is used for reproductive pheromones, social odors, warning signals, and a host of other smelly signals.

Returning to our mice and their urine, another group of researchers, headed by Dr. Jane Hurst at the University of Liverpool, has looked beyond the MHC to another possible source of individual recognition. These odorous molecules are called major urinary proteins (or MUPs). The genes that encode for the MUPs are as polymorphic as the MHC group and code for proteins that bind small volatile molecules found in urine. The MUPs then slowly release the odor molecules in dried urine providing a long lasting and identifiable signal. Dr. Hurst and her coworkers have shown that a single mouse can have up to 7–12 different MUPs in their urine presumably based on genetic differences.

Mice are social animals and often mark territories with urine signals. On the borders of these territories, the mice often engage in a sort of chemical signal one-upsmanship. Intruders leave their own scent marks on the original, and the mouse in the home territory responds with a subsequent countermark. At this point, the mice are engaged in a veritable "pissing match" trying to cover their territory with their own marks, like dogs and fire hydrants. Interestingly, if the intruder is a experimentally manipulated brother of the original mouse (has a different set of MUPs), the counter marking is intense and aggressive. In this situation, the manipulated brother has a similar set of MHC genes as the original animal, but a different set of MUPs. Conversely, if a new intruder is introduced that is a brother who is not manipulated, same MHC and MUP, the counter marking is not worth noting. This work appears to point to a second group of genetic products that are critical to individual recognition. Regardless of whether the MHC or the MUP is the scent of choice, both provide the mouse with the ability to essentially "smell" the DNA of another animal and identify individuals.

4.5 Hey, Soul Sister

I really loved my time in Woods Hole, Massachusetts, and the small community really had accepted my wife and I as members. The Cape is a truly wonderful location, both from an environmental and scientific point of view, but the people are relatively slow to warm up to outsiders. Though, once you are a part of their community, there is a sense of lifelong loyalty that comes with the patience to win Cape Codders over. After my Ph.D. work in Cape Cod, I took a postdoctoral position as a researcher at the University of Colorado Health Science Center in Denver, Colorado. I had visited Denver in the past during some family vacations, but moving across the country was a completely new endeavor. Our move to Denver was marked with an early visit to the auto repair shop. Having moved in August, Denver was quite warm that day and the waiting room for the repair shop was not air conditioned. As I sat patiently, reading some magazine, an older gentleman came in after dropping his car off in the garage. With a quick glance in my direction, he offered to buy me a pop because he thought I needed a little refreshment. I politely accepted and he quickly noticed by my accent that I was not from Denver or even Colorado. This led into a rather pleasant conversation about the virtues of Colorado. In my 4 years in Cape Cod, I never experienced a stranger offering a pop (or soda in that part of the country) to me. Again, I love Cape Cod and the people in the Cape, but there is a different type of friendliness that is inherent to the Cape as opposed to Denver.

Now that I have settled in the Midwest of the United States, I am lucky enough that graduate students from all over the country (and some parts of the globe) come to work in my lab. The diversity of ideas, backgrounds, dress, and language make for interesting conversations during lab meetings and informal gatherings. What becomes readily apparent rather quickly in our conversations is the difference in words and dialects. Students from the South as opposed to the west or Midwest have the characteristic southern accent and drawl. Those from the Northeast have either a strong New England or Bostonian accent. Some students come in with such strong accents that a short explanation may be needed to understand what each of us is saying.

Even the lab's research animal, the crayfish, has numerous names across the country. Although the official scientific name is crayfish, the heart of the United States (Missouri, Oklahoma, Kansas, Nebraska, and Arkansas), as well as the west coast (California and Oregon) refers to our research animal as crawdads. The Deep South, the Atlantic coast, and some of the plains areas call these creatures crawfish. The upper Midwest and Northeast refer to them as crayfish. Finally, my Australian postdoc referred to the animals as yabbies. Thus, during our conversations, the listener can tell a little about the homeland of the speaker by the use of certain words as well as the dialect. Similar distinctive word choice can be done with the soda, pop, and coke triumvirate as well as the "you guys/you/y'all/you all" usage for a group of people.

Unlike the MHC recognition above (which is genetic), the type of friendly nature, the dialect spoken or word choice is a learned phenomenon based on the local environment in which the speaker grew up. These cultural and regional distinctions in language use or dialects are badges that serve to associate the individual with different groups like New Englander or a Southerner. The familiar pop or soda usage determines if someone is "from around here" or the potential "in" group versus a visitor from a different part of the country. Animals also have learned badges that allow them to distinguish neighbor from visitor or in the animal kingdom, friend from foe.

The paper wasp, *Polistes fuscatus*, builds large nests composed of "paper" that have a large number of individual cells hanging downward. The wasps need bits and pieces of wood and are usually found around woodlands. They chew on the wood and produce a paste mixture from which they build their nests. The female wasp that starts this process is called the foundress and she is the dominant wasp in this nest. The foundress begins to lay eggs in each of the chambers and as the eggs begin to grow and hatch, she continues to build other chambers for more eggs. This process continues on and on until the first eggs begin to hatch.

Once multiple wasps have hatched (all females at this point), some of the sisters and even the foundress can fly off to find other nests in which to build cooperative broods. Those sisters that stay on the original nest will help build more egg cases, defend the nest from intruders, and help forage for caterpillars. Upon the return from foraging flights, the sisters will share bits of masticated caterpillar with the larvae and with each other. Herein lies the problem for the paper wasps, how does the wasp know that a particular sister is returning with food or is an intruder?

Dr. George Gamboa has been studying paper wasps for 30 years and began to notice that paper wasps tolerated the presence of nestmate sisters more than non-nestmate sisters. This was true whether the encounters took place in the field or on the umbrella like nest. The friendly treatment also extended to the brood. Foundress wasps either ate or bit larvae and eggs from unfamiliar nests and treated larvae and eggs from their home nests with a far more gentle approach. Having carefully controlled for relatedness, the wasps did not appear to use a measure of genetic relatedness as the source of recognition. Unlike the MHC recognition outlined earlier, this

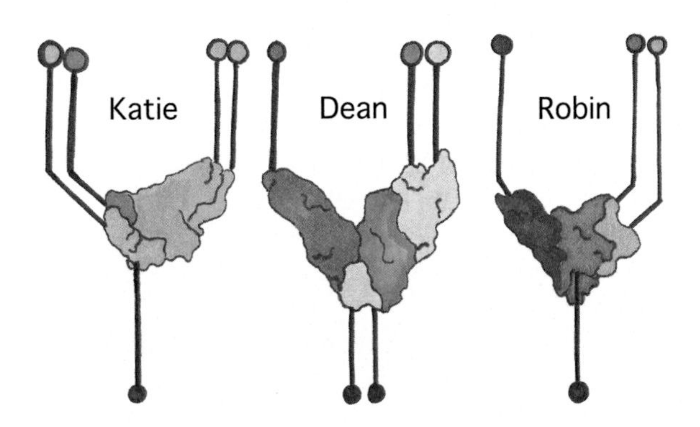

Fig. 4.4 MHC molecule

type of recognition appeared to be more related to the local environment or in other words, the nest itself (Fig. 4.4).

In a similar fashion to the salmon homing described in the next chapter, paper wasps "imprint" to the local food and nest smells and use this smell to recognize nestmates from non-nestmates. In a series of ingenious experiments, Dr. Gamboa and his students raised colonies of paper wasps from eggs to adulthood under identical conditions in his laboratory. The wasps were exposed to identical food throughout their life, identical environmental conditions, and nest making materials. Wasps exposed to other colonies were treated as nestmate sisters by the foreign colony. In addition, if the female wasps from different colonies were exposed to fragments of the same nest, the two different wasps treated each other as nestmates. It turns out that there are nest-specific chemicals that are "woven" and painted onto the nest by the queen as she strokes her abdomen on different parts of the nest. If the nest or other wasps are washed with a special solvent that removes these chemicals, the recognition is gone and the wasps are treated as intruders. Since the queen creates the initial nest, lays the initial eggs, and paints the nest with her chemical cues, the recognition signal is a combination of local environment interacting with the genetics of the queen.

Across college campuses, Greek life is a big part of the culture. Sororities and fraternities are easily marked by their three characteristic letters such as Delta Tau Chi. Sweaters, sweatshirts, hats, and bags all mark the wearers as sisters or brothers of a certain Greek group. Even off campus and later in life, if two people recognize that they share the same sorority bond (even across different campuses), they are sisters for life. For sorority sisters, there is an increased level of friendliness and a willingness to immediately help if needed. Three letters, say ΔTX, are all that are needed to signal a group. Within the paper wasps society, the chemicals that queens place on nests serve the same function as saying "soda" (indicating a New Englander) or crawfish (indicating a deep south location) or wearing a sweater with three Greek letters on the front. Paper wasps with the same nest and queen smell are sisters for life.

4.6 Are You My Mother?

Nestmate recognition is critical for maintaining and controlling important resources. Paper wasps invest a lot of energy and time into making their nest and the queen has invested enormous energy stores into producing eggs. Protecting that investment into their reproductive efforts through nestmate recognition provides the paper wasp an evolutionary advantage. In some bird species, the recognition of offspring or relatedness is not as fully developed and some bird species have evolved strategies to take advantage of this reproductive loophole. Cuckoos are one such type of species. Cuckoos will lay their eggs in the nests of other birds and then leave the parenting skills to the very accommodating host bird. This type of system is called brood parasitism, and the cuckoos are particularly adept at this lifestyle (Fig. 4.5).

Fig. 4.5 Cuckoo egg in nest

The cuckoo and another brood parasite, the cowbird, have evolved egg mimics. This adaptation is where the cuckoo or cowbird egg looks very similar to the host egg. The reed warbler is a favorite target of the common cuckoo. The warbler's egg is tan in color with darker colored speckles. Although the cuckoo's egg is about 50 % larger than that of the warbler, the cuckoo's egg is almost identical in color. The eggs are close enough in color that the host bird, in this case the warbler, raises the offspring as its own. There is a significant cost to the warbler in raising the cuckoo's offspring and often, the cuckoo is provided food at the detriment of the warbler's own offspring. Given the reproductive cost to the warbler, there is evolutionary selective pressure for recognition of its offspring. Thus, warblers that are more successful at detecting and rejecting cuckoo eggs will have a higher reproductive success rate than those warblers that are blind to the cuckoo eggs. Of course, there is selective pressure on the cuckoo also. If the cuckoo's eggs are significantly different in color from the warbler's eggs, then there is an increased chance of recognition of the cuckoo's egg which is often followed by rejection. The increased selective pressure for warblers to recognize eggs and the increased selective pressure for the cuckoos to decrease this recognition is termed an evolutionary arms race (as a reference to the cold war era arms race). Recognition of one's offspring by the parents is a critical aspect of reproductive success.

One of the parental hallmarks of mammals (and bird species) is the concept of extended and intense offspring care. Mammalian (including human) young are incapable of taking care of themselves, and thus, parental nourishment, protection, and learning are important for the successful rearing of offspring. The reproductive consequence of long periods of nurturing by mammalian parents is the loss of repeated and large scale reproductive efforts. In contrast, many invertebrate species have very little parental care and can invest more in large numbers of offspring in the hope that a small percentage of them survive to adulthood. The German cockroach is rather small from a cockroach perspective, but are champions when considering their reproductive efforts. A single female can produce over 10,000 offspring within a year. For cockroaches, the loss of a few hundred offspring is a common occurrence. For mammals, the loss of a single offspring is something of great concern.

Given the importance of rearing the few, but precious young, mammals have evolved a strong system for the recognition of their offspring, and this system is based on chemical signals. Rabbit pups are able to recognize and prefer their own nest over that of a different mother's nest as early as a single week after their birth. Similar work has shown that mice exhibit the same types of preferences where offspring are drawn to the odors of their own related mother as opposed to another unrelated mother. Returning to the work of Yamazaki and Beauchamp covered previously, this recognition is probably through the MHC complex, but with the parent offspring recognition, the offspring learn the appropriate odors that are associated with maternal recognition as opposed to some innate recognition.

As important as recognizing ones mother is, recognizing that an odor is coming from a dangerous parent can be just as important in the mammalian world. Infanticide is a phenomenon where an adult male or female within a social setting kills an offspring of the same species. This behavior, although not common, isn't rare in the animal kingdom and has been documented in a number of invertebrates, birds and mammals. A deeper explanation for the presence of this behavior across a number of species is beyond the scope of this book, but one of the main theories for the evolution of infanticide involves sexual conflict. In a number of mammalian societies, a single male often has access to a larger number of females for reproduction. Within these societies, this male (alpha or dominant male) spends a very short period of time as the top dog. If females within his group are currently caring for offspring from the previous alpha male, his reproductive success can be limited. Thus, the alpha male will frequently kill the offspring of his previous competitor which will allow the females to become fertile again. For mammalian offspring, the close presence of an unrelated male is often a sign of life threatening danger.

Infanticide is present within mice communities, and mice pups will produce more ultrasonic calls to mother mice in the presence of odor from male mice that have performed infanticide as opposed to male mice who have not killed any offspring. These pups will also attempt to move away from these infanticidal males, but will remain stationary to their own fathers. Although the specific chemical cue has not yet been found, there are odors produced by males who have killed offspring. Pups are sensitive to this odor that is missing from their own fathers or mothers. Presumably, the offspring are seeking the safety that their own parental odors represent.

4.7 That New Baby Smell

The offspring parent bond and recognition is as important within our own society as in animal societies. In many of the animal studies, researchers have focused on the chemical cues given off by either parent. These cues help control the offspring's behavior, mood, and overall health. The aromas of parents can promote a sense of calm, feeding, and reduce stress. More recent work has begun to demonstrate that the odors that babies produce (beyond their diapers!) could have significant and important impacts for parents even if the parents are unaware of these effects.

Psychologists have recognized for years that the bonding process between baby and mother is critically important for the health and survival of the baby. Work on bonding between babies and fathers started a little later than work on mothers, but has demonstrated the importance of this bond also. Bonding, as a psychological process, draws the parent closer to their baby and results in a much stronger physical and emotional attachment. As mothers and fathers cuddle or hold their baby, they may be doing something far more subtle than just providing a loving physical environment for their new child.

An interesting study, led by Drs. Johan Lundström (at the Monell Chemical Senses Center) and Thomas Hummel (Universität Wien), found that new mothers showed an increased neurological response when presented with infant odors as compared to women without children. The response was neurological rather than psychological because both groups of women gave the same verbal ratings for the odor qualities. The brain responses were measured by a machine called a Tesla MRi-scanner. The scanner measures both blood flow and oxygen concentration in different areas of the brain and areas light up (have increased oxygen and blood flow) as that part of the brain is being used. The women essentially had puffs of odor delivered to their nose that were derived from cotton undershirts worn by the infants for 2 days. Both the new mothers and non-mother women rated the odors equally for pleasantness, intensity, and familiarity. In addition, both groups, independent of whether they had offspring or not, showed more activity in the area of the brain (neostriate) that would cause them to care for the infant. The neostriate is associated with planning and movement. Thus, the possibility exists that odorant molecules in babies trigger an unconscious response to care for and about the health of helpless infants.

Furthermore, the new mothers showed increased brain responses in a different area of the brain (dorsal caudate nucleus). This part of the brain is associated with a number of cognitive functions such as learning, memory, emotions, and language. Finally, both of these areas of the brain (neostriate and dorsal caudate nucleus) have a preponderance of dopaminergic neurons. These neurons release the neurotransmitter called dopamine which is link to the reward mechanism in our brains. For example, if you are with a group of friends and tell a funny joke, the resulting laughter triggers little tiny jolt of dopamine and you feel some pleasure or happiness. This often promotes you telling other jokes just to receive additional jolts of dopamine. This cycle, dopamine to pleasure to seeking more dopamine, is a positive feedback loop that can lead to addiction. In this case, the odor may serve to find a good caretaker for the baby and to increase the bond between a mother and her baby. The fact that new mothers have an increased response in some areas of the brain to infant odor may further enhance this bond between the mother and child as opposed to a bond between a woman and unrelated child. In a way, these infant odors, which allow mothers to identify their babies, are sneaking their way

Fig. 4.6 Wispy baby rewiring brain

inside the brain and influencing the pleasure and reward system of the brain without the mother becoming consciously aware of that connection (Fig. 4.6).

In this chapter, I have written about the different identities that people and animals have and the roles that they play on a daily basis. Whether there is a need to recognize self from nonself in order to destroy dangerous intruders in our bodies or the need to recognize those in a unique or special social role such as our friends, our family, or our clan, chemical signals are there as sure as name tags and Greek letters. Some of these chemical badges are inherent to who we are because they are generated by the very DNA that determines our very being. These signals surely represent us to the world in a revealing way. Although we can change our clothes or hair color and we can attempt to cover our odor tracks through shampoos, deodorants, and perfumes, the chemical signals generated by the MHC complex within our bodies announce our DNA to anyone astute enough to pay attention to it.

Chapter 5
Home Sweet Home

Finding Shelter and Safe Spots

The clock in my office reads 5:15 pm and I start to pack my messenger bag with the work I want to take home. A truly productive day has come to an end and I will be glad to head home for some good down time. Since rain has been present most of the day, I decided to drive rather than walk to work today. After a day of teaching, writing, and thinking about research, my mind is temporarily drained of any real deep conscious thought. I feel as if my mind is on autopilot as I shut my office down and head down the hall to the stairwell. Step after step, I am guided toward the doors, then the stairs, and then the parking lot with my car as if drawn by some invisible force. As I reach my car, I plop my bag down in the passenger seat and start the engine. Putting the car into reverse, I back the car out of the parking spot and begin my journey home.

The focus of this chapter is around homes and shelters and how animals select a good home and then return to their home. For many animals, home provides a safe haven for sleep and raising young. Just like most of our daily sojourns out of our homes to perform jobs or gather staples, animals need to leave the safety of their shelter to forage for food or supplies. Bees will travel great distances to find pollen and nectar to make their sweet honey. Ants trek much farther distances than the bees to gather leaves, plant matter, and even other insects as critical staples for their nest mates. These foraging trips can be broken down into two simple behavioral tasks. First, there is the trip out from the home which can entail either a guided expedition to a set destination or an exploration in search of treasure. Bees will perform an elaborate dance, called a waggle dance, which communicates the fairly precise location of a favorable patch of nectar and pollen. The outward foraging trip of the bees that have watched the dance is guided by the directional information contained within the elaborate maneuvers of the dance. Ants will lay down chemical trails when returning from a successful foraging trip. As subsequent ants make their own foray out into the world, they will use this chemical highway to guide their way to the resource rich patch and then back to their home nest.

Second, there is the need to relocate the shelter or home after going some distance away. As I ease my car out onto the road from the parking lot, I use my own set of cues to find my way home. I am not actively thinking about how to get home as I turn left

P.A. Moore, *The Hidden Power of Smell*, DOI 10.1007/978-3-319-15651-4_5

at the first stop sign. This route home is so engrained that I need little conscious thought to make my way. Glancing left and right as I drive, I do perceive subtle signs that I am on the right path. To my right is a rental house with a black metal staircase on the outside that serves as an entrance to the second floor apartment. I pass this house on every trip to and from work and over the course of two decades; I have had the chance to memorize just about every external detail of the house. This rental house serves as a visual landmark that reminds me that I am on the correct path. Further along this road, I reach a four-way stop that requires a right hand turn. After briefly stopping and noting no other cars at the intersection, I turn right and continue my journey. After a short distance, I cross a set of railroad tracks. The click-clack of the tires crossing the train tracks is an auditory cue telling my brain that I am on the correct path home. Although not as prominent in our conscious thought, these sounds are subtle messages that provide supporting information to the visual landscapes that we are on the right path. After a couple more turns, I arrive at my house.

How do I know that this is my house? Although this seems like quite a ridiculous question, the consequences of constantly walking into the wrong house could be dire for me, and for animals, entering the wrong burrow, den, or nest could be deadly. As I pull up to my driveway, I can easily recognize all of the visual elements that comprise my house. The reddish brick, the red and yellow painted trim features, the Forsythia bush on the left, and the Spirea on the right all signal to me that I have entered the correct driveway. Further evidence that I have arrived at the correct residence greets me as I open the door to my home and cross the threshold. I am met with a host of smells that are unique to my home. Currently, my home houses a dog, two cats, two teenagers, my wife, and myself. Each of us contribute a set of natural aromas due to our gender, age, species, and diet and a set of artificial fragrances that arise from the assortment of soaps, perfumes, deodorants, and shampoos that we all use. As I close the door behind me, I stop to take a very deep breath and inhale this chaotic symphony of odors. This is home and I recognize this mixture of fragrances as my sanctuary.

In addition to finding their way back home, animals need to find a suitable spot to make a home. Animals' homes have as many names as there are types of animals. Shelter, den, nest, cave, and burrow are just a few names used among scientists and the general public. All of these locations serve multiple purposes for animals. One of the primary functions of a home is protection from predation. A good shelter lives up to that name by providing refuge that either hides the animal from predators or offers physical protection by being underground, high in a tree, or under some hard objects like rocks. Homes can provide protection from the elements such as wind, rain, cold or heat, and places in which offspring can be reared due to close proximity to water or food.

5.1 You Can Never Truly Go Home Again

The odors that greet me when I step into my house immediately assure me that I am home. I can recognize all the different rooms and their unique set of odors. Walking from the kitchen with the aroma of a home cooked meal (spaghetti with garlic bread last night) is different than the living room with its mixture of white rain and vanilla candles in the corner. The bedrooms upstairs each have their subtle "shades" of odors due to the concoction of age and sex dependent body odors and the myriad amount of shampoos, hand creams, deodorants, and perfumes that the occupant chooses on a daily basis. When I travel to my parents' house, a different suite of odors are present inside my childhood home. Although familiar in a distant sort of way, the odors do not signal my house. Having last lived in this place several decades ago, the aromas are far more different than the visual landscape that is present.

The association of home with a location shifts with the odor landscape. Imagine a nice dining room table all cleared so that we can perform our experiment. On top of this table, a smattering of close relatives have each brought a towel and neatly laid them out on the table. I say close relatives because for this experiment, I would want the tester to be familiar with and exposed to the "home" odors of the different towels. For me, I would have one from my house, my parents, and my brother freshly washed and dried just for our trial. A small sample indeed, but enough for this thought experiment. After donning a blindfold, the tester would grasp a towel and bring it to their nose. After a brief exhale to empty their lungs, they would take a hearty inhale to fully experience the euphonious aromas of the towel. Could the tester tell who the owner of the towel was? I am not a professionally trained perfumer nor even an instinctual odor expert like my dogs, Cedric or Loki, but I would like to imagine that I could indeed recognize my home towels from those of my parents or my brother. I would think that the daily exposure to the set of chemical compounds floating in the air of my house would trigger some odor recognition. This sniff may even induce an odor memory where I am immediately swept (mentally) back to my house.

As mentioned in Chap. 1, during the summer time, I spend an intense field season at the University of Michigan Biological Station in the upper part of the Lower Peninsula in Michigan. While there, I teach and do research with my graduate students and undergraduates. The faculty stay on campus in small, but cozy cabins on the shores of Douglas Lake. I have a set of bedding, sheets, blankets, and comforter, which are set aside just for this yearly trek north. During the winter time, I store this bedding in plastic bins in my basement. Every summer, those sheets and comforter capture the fragrances of northern Michigan wooden cabins, lake side breezes, and pine trees. In the spring, when I begin my yearly packing, I can open those bins and be instantly transported in time and space to a summer on Douglas Lake. I recognize these sets of odors as a different type of home. Different than my winter dwelling and different from my parents' house, the smell of that comforter still creates an image of home for me. Traveling "home," whether to my house, my parents' house, or northern Michigan, is a relatively easy trek. I can read maps and follow traffic signs to get me to my destination. Though my eyes tell me I have arrived at my desired location, it is the subtle odors of that particular place that confirm I am home.

Some animals do not have the luxury of visual road maps or sign posts to guide them home from either short or great distances. Once a home is found, virtually every animal will have the need to leave and return to the specific location of their home. These trips are such necessary behavioral tasks as foraging runs, searching for potential mates, scouting the neighborhood for new members, or security patrols to look out for potential threats. These trips can be as short as a few millimeters or as long as halfway around the globe. One of the champion treks guided primarily by olfactory cues has to be the long and arduous task faced by salmon returning to their natal streams.

A fairly common and well-known story, these fish have an uncanny ability to return to the place of their birth even after being away for 6 years. Given the economic importance of salmon across the globe, the ability of these fish to return to their exact stream in which they were born and spent their juvenile years after years away is both legendary and well documented from a scientific point of view. As an adult, when I walk into my parents' house where I spent my juvenile years, I notice that the odor signature is very different than I remember. The use of spices has changed, the characteristic smell of the laundry soap has been altered, and a few different unfamiliar candles provide new aromas. I travel home for periodic visits to say hello or get caught up on the latest family information, whereas salmon make the odyssey to their place of birth in order to reproduce. In contrast to my short visits, these salmon are indeed trying to go home again. What is essential, at least from the salmon's point of view, is that their birth place needs to smell identical (or at least similar enough) to when they left that home many years earlier. Salmon have an interesting life cycle as they are an anadromous species of fish. Anadromous fish are born and live their juvenile lives in freshwater bodies. Once they reach a certain age, salmon migrate out to saltwater to life as a marine species. When the urge to reproduce hit them, they all begin the long trek home. It is during this return trip that salmon need that alluring scent of home to guide them to their natal stream.

First documented by Norwegian fisherman in the 1600s, the remarkable journey has been of intense interest ever since that time. Leading the discovery of this long distant olfactory trek is Dr. Ole Stabell in Norway and Dr. Thomas Quinn in the United States. Olfactory imprinting is thought to be the underlying mechanism at work with salmon. Imprinting is most widely known in mammals and birds and happens when a newly hatched or birthed offspring imprints a concept of "mother" or "safety" on one of the first things the offspring interacts with. The concept was developed by the Nobel Prize winning ethologist, Konrad Lorenz, who showed that geese hatched in an incubator will nearly imprint on anything that is moving within the vicinity. Lorenz noticed imprinting only occurred during a specific period of time which he called the "critical period." The behavioral phenomenon is fairly well demonstrated by geese imprinting on Lorenz himself, and any search through publicly available videos will show ducks, geese, dogs, and cats imprinting on various people and even toys. The salmon also have a critical period for olfactory imprinting. The exact time period of imprinting is under debate, but regardless of time period the behavioral outcome is the same. During the early stages of their life salmon live within streams, and each stream has a characteristic suite of chemicals that make up their home stream odor.

Standing on the edge of a headwater stream in northern Michigan (or in Washington or in Norway), rivers look relatively identical. The slope of the flowing water may be different, and Norway certainly has different types of plants as compared to Michigan or Washington, but to the salmon, each home stream has a distinctive set of odors that signal "home" just as much as our mix of candles, foods, and perfumes tells us we are home. The "perfume" for the salmon's home stream is an interesting mixture of chemicals coming from both the biology of the stream and surrounding area as well as the salmon themselves. Headwater streams or rivers are bodies of water that essentially are the first step in rivers as they travel from their sources to the sea or lake. In temperate forest regions, these streams are surrounded by hardwood forests. Typical in most forest regions, there are often leaves from the previous year's growth littering the floor of the forest. In addition to the leaves, the mixture of tree species exudes from their roots certain organic chemicals as part of their natural process of photosynthesis and respiration. All of this organic matter (leaves, roots, soil animals, and organic exudates from the trees) act as "coffee grounds" for rain and snow melt to flow through. As rain falls, the water that gets absorbed into the soil percolates through all of this organic matter and eventually flows into the river as ground water. During the trip from cloud to ground to river, the rain water picks up the flavors of the forest and delivers these flavors to the salmon's nose just like coffee flavors our taste buds. This process is almost identical to how tap water dripping from the coffee maker's reservoir through the coffee grounds into our mugs delivers a wonderful morning wake up aroma to our noses. Water flowing through coffee grounds becomes delicious coffee. Just as changing the roast of the coffee, say from an American light roast to a French dark roast will significantly alter the flavor of the coffee, alterations in the structure of the forest will change the organic compounds in the river water. If for example, a stand of oak trees die and are replaced by a group of white pines, the organic composition of the percolated rain water will also change, changing the smell of the home stream.

The water saturated with the smell of the forest floor mixes with the body odors of the salmon and other inhabitants of the river to produce the unique signature to which the salmon imprint or learn. This odor, the natal or home stream, guides the salmon during the last phases of its journey from the far reaches of the ocean back to its place of birth. The entire journey isn't guided by the sense of smell. In the middle of the ocean, salmon rely on cues or information from their other senses. There is evidence that salmon may use the magnetic field of the earth in combination with visual cues to allow them to find the coastline near their natal stream, but once there, olfactory cues are the primary source of information for salmon needing to find their way home highlighting the role of olfaction to help these fish reach their home stream.

Whether the odor is in air or water, the chemicals form an odor plume. I presented an analogy of a smoke stack in order to visualize the movement of chemicals by wind or water. With that visualization in mind, imagine the southern coast of Alaska or the western coast of Norway. Along each of these coasts are hundreds of streams, rivers, creeks, and wetlands all releasing their respective odor plumes into the ocean. All of these river and stream plumes would intermingle and mix along the coastline making the specific determination of a location of a single stream a very difficult task. Now imagine a different scenario. I take you into a large open field. At one end of the field is a series of 10 tables, and each table has a single candle placed upon it. Each candle has a unique aroma where one candle smells exactly like the inside of your house and the other nine smell like your neighbor's houses. Hopefully, you have clean neighbors. Finally, behind each candle I place a large fan to help deliver the odors down the field. You and I now walk to the other end of the field. I will give you a couple of minutes to smell the different aromas, but then I blindfold you. Your task? Find your way home using the scented candle. This is essentially the task that the salmon are faced with.

Thankfully, the salmon has a little advantage being an aquatic organism. The main difference between freshwater and saltwater is the presence of salt. Although this last statement is fairly obvious, the implications of this simple fact are rather large for the salmon story. The salt increases the density of oceanic water, such that as a river empties out into the ocean, the water from the river actually "floats" on top of the oceanic water for quite some distance. Eventually, the river and oceanic water mixes and becomes indistinguishable. Before these waters mix, each of the different rivers has a slightly different density and as they all flow together into the ocean, they form different and distinct layers of river water. This is somewhat akin to a stack of different flavored pancakes that one would have for breakfast. Pumpkin pancake on top, followed by a raspberry pancake, then blueberry, a wheat pancake, and finally, a chocolate chip pancake on the bottom. Each different flavor of pancake would represent a different stream flowing from the coast of Alaska into the Pacific Ocean. The salmon use celestial or magnetic cues to get close to the coast and once there, they swim up and down in large vertical forays supposedly sampling all of the different pancakes of river water. After finding their preferred flavor of pancake (the flavor of home), they follow their nose back to their natal stream.

5.2 The Long and Winding Road

One of the ways that I attempt to stay healthy is through cycling. In many ways, the roads around my home town are ideal for cycling. They are rarely traveled and given the lack of any significant rise in elevation; the roads can travel in a straight line for miles and miles. (Although I would prefer the different terrains and challenges that mountain passes or even hills would present, I make do with where I am). Cycling provides me with two different forms of release. First, the rhythmic pedaling offers me a very good aerobic workout without the pounding on my older knees and poor ankles from running that leaves me quite sore the next day. Second, because of the lack of elevation changes and car traffic, I can almost approach a meditative state where I am aware of the road and potential hazards, but am also focused inward on my thinking. I do not have to really pay much attention to the path that I take having traveled these roads enough to notice the landmarks to tell me where I am. I have a mental list of the key landmarks around my various routes. I tend to head out into a headwind so that I have a tail wind on my return trip when my legs begin to feel heavy. On my southwest route, I know there is an old abandoned high school at the corner of these two very long and straight roads. Going out straight west, I hit a river and a fairly large stable. If I were ever to forget these landmarks and truly get lost, I can always count on seeing one of two yellowish tan water towers and ride as straight as possible toward those towers on the edge of town.

During their travels away from the safe shelter of their burrows or dens, animals use different mechanisms to locate their way back to their home. When I use the water towers to ride home, I am performing a behavioral task called visual landmark orientation. The water tower serves as a landmark and mentally, I have constructed a rudimentary map of the location of my home in relation to the water tower. Animals can construct a map in similar fashion to what I do with my house and the water tower. They could also simply navigate toward the landmark and perform a standard search pattern until they run into their home. In many ways, I am quite lucky on my rides because someone else (in this case the DOT of the state of Ohio) has laid down roads and highways that I can follow. In addition, at the intersection of two roads, the DOT has placed nice signs that label the roads and allow me to mentally construct a map that tells me exactly where I am. I could simply keep track of which roads and turns I took to head out away from home and reverse that track to find my way home. This method is not necessarily the most efficient way home, but it works fairly well.

Some animals have their own highways and roadways with very specific road signs that help guide them to and from their home base. One of the most studied systems of highways is the odor trails used by ants on their long distant foraging trips. First worked out by the famous Harvard biologist E.O. Wilson, the chemically based highways are a marvel of simplistic and ephemeral engineering. Ants make large-scale foraging trips to locate distant food sources and need to return to the nest with the captured food. In many cases, the food sources are often quite rich and require several members of the colony to assist in gathering this food. For humans, traveling to a local grocery store is a relatively easy task. If we have lived in a location for even a short period of time, we memorize the route rather quickly and the route doesn't change (unless the store closes up or moves). For ants, their trips are made across the complex environments made of shifting landscapes of grass, sand, branches, and plants. In addition, once the cache of food is completely depleted, any pathway back to that location is useless to the ants. Here is the quandary for the ant "engineers" that want to build a highway to the grocery store. On one hand, the colony needs a reliable highway that can be followed by numerous members of the colony. On the other hand, the highway needs to be dismantled once their "grocery" store has sold out of their goodies. A simple and elegant solution is a highway made out of pheromones.

Within the ant society, the workers are the ones that leave the nest in search of food. Harvester ants (*Pogonomyrmex barbatus*) make foraging trips in search of seeds or mushrooms to store within nest granaries. These foraging excursions can involve traveling as far as 20 meters from the nest and take as long as an hour to locate suitable food. For ants that are roughly 10 millimeters in size (about the size of two capital "I"s stacked on top of each other), this translates into a trip of 2000 body lengths. In human terms, we would walk over 2 miles just to fetch some milk and eggs. On their journey, the ants will periodically touch their abdomen to the ground and leave a little drop of a trail pheromone. As the ants continue on their

Fig. 5.1 Ant highway

walk, their trail production leaves a dotted line of pheromones across the ground of their native habitat. In this way, each foraging ant constructs an outbound pheromone highway leading toward a seed or small mushroom. On the return to the nest, the same behavioral pattern is repeated producing an even stronger scent trail. As other harvester ants encounter the trail, they will detect the presence of a signal by touching their antennae to the ground where the trail pheromone has been deposited. Once this new ant recognizes the trail signature, they too will take off toward the food source depositing even more pheromone on top of the current trail. Each ant becomes a new chemical engineer adding to the strength and robust nature of the current pathway (Fig. 5.1).

The nature of the chemical trail is such that the signal evaporates rather quickly. Studies have shown that the trail exists around 2 minutes if not reinforced by new pheromone from traveling ants. The volatility of the scent pheromone allows the trail to remain rather narrow which keeps the ants focused on the current trail to the food source. If the signal was less volatile or more constant, then the pheromone trail would slowly widen and increase the potential for ants to take a wrong turn off of the trail. Occasionally, the ants do get lost, but amazingly other nest mates will help guide these lost foragers back onto the trail. A good rich food source will produce an ant superhighway that rivals any rush hour traffic in a big city. Fortunately for the ants, their highway construction process and "driving" abilities are far superior to ours, so there are rarely any backups or slowdowns on their trails.

Once the food source starts to diminish in quality, fewer and fewer ants return to that location. If the food becomes depleted, then there is no need for the ants to use this particular chemical highway. If foraging ants were to run into this older trail and attempt to locate an empty food cache, the ants would be making wasted foraging attempts. At this point, the volatility of the trail pheromone causes the pheromone trail to dissipate into the wind. Unused trails disappear from the landscape within 2 minutes, and there is no trace of the old highway. The ephemeral nature of these highways creates a dynamic chemical architecture from the ant's point of view. On a daily basis, new trails are constantly being laid down by successful foraging ants and the old depleted highways vanish from the landscape. If we had this level of efficiency in construction and destruction of our roads, traffic problems might disappear as quickly as unused ant trails.

5.3 Where I Lay My Head Is Home

In the marine environment, a large group of organisms lead sessile lives. This means that, as adults, these organisms are permanently attached to the substrate and will live their entire adult life in that one spot. So, choosing the correct spot to become rooted to can literally be a life and death decision. Organisms like barnacles, corals, sea anemones, mussels, and sponges all lead sessile lives as adults. A good location for these organisms can provide all of the items outlined above for a good home. Animals need to have access to valuable resources such as plenty of food, protection from predators, and finally, other organisms of the same species for potential mating opportunities. Even some mobile marine organisms are so slow moving that the choice of a good habitat is critical. Some of the most beautiful animals in the oceans, marine gastropods, fall into this category of slow moving habitat seeking group. So, if an organism is sessile or extremely slow moving, how can critical habitat selection occur?

Although the adult form of these animals is slow or sessile, the juvenile form is actually quite mobile. Many marine gastropods have a larval life stage called a veliger, and at this stage they are released into the water column to freely swim around or be swept along by different ocean currents. The veliger looks very different from the adult and is adapted to do two things. First, veligers have special hairs on their body that allows them to perform a rudimentary swimming behavior. They are not good enough swimmers to actually move against currents, but they can certainly move themselves up and down the water column. This swimming ability is enough to enable them to perform their second function and that is to locate a new home based on the sense of smell. With a special "'nose" called an apical sensory organ

Dr. Mike Hadfield at the University of Hawaii has made a career out of studying these little creatures, how they know when a good home is present, and the amazing transformation that occurs once that home is found. The life cycle of these veligers is fairly well established now thanks to the long and hard work of Dr. Hadfield and his colleagues. Although, the exact timing and details are different for each species, in general the veliger story essentially is as follows. The adult gastropods (the sea slug *Phestilla sibogae*, for example) have external fertilization which means that male and female gametes are released into the water column to come together to become a fertilized egg. The eggs float in the water column for a period of time on the scale of days to weeks. Once the eggs hatch, the veliger larvae begin their travels on the ocean currents. At this point in time, the veliger has a clock that begins to count down. This clock is called "competency" and essentially means that there is a window of opportunity for this veliger to find a good spot to call home. If the clock runs out, akin to the stroke of midnight for Cinderella, the veliger will die without metamorphosing into the adult form. This is a forced choice for the veliger due to this competency clock: find a good home or die. The clock is called a competency clock because during this period of time, the veliger has the ability (competency) to metamorphose into an adult form, whether crab, snail, or some other benthic creature.

During the swimming veliger phase, the downward swimming motion is triggered and controlled by a tiny nervous system connected to the veliger's apical sensory organ. This organ, the nose, is sensitive to specific chemicals that signal "This is a good home, swim down here!" Just as I walked into my home and recognized the wonderful aromas of my home, the veliger travels the ocean sniffing its way to a new safe haven. Without the right chemical or mix of chemicals, the veliger continues its journey in search of a new home. Depending upon the species in question, the chemical can be a sign from fellow species members, potential food items, or even neighbors.

Compared to its fellow sea slugs, the sea slug, *P. sibogae*, is a fairly drab organism. Roughly 3–4 centimeters in length, the mostly white slug has a couple of tentacles near the head that function to sense chemicals in the environment. What is unique about this species is that *P. sibogae* is entirely dependent upon corals for its food. Without a good healthy coral reef to consume, this animal would perish. The veliger of this slug has a critical decision to make in regard to a new home and that decision is, "Is there enough food here to last a lifetime?" While the veliger is idly floating along the ocean's currents, its nose is on high alert for that suite of chemicals that signals a nice juicy and healthy coral reef. That suite of chemicals are metabolites coming off of *P. sibogae's* only prey the coral, *Porites compressa*. These corals, as drab as the sea slug, grow as lumpy boulders in shallow water or as fat fingers in deeper waters. In either form, the corals grow in very large colonies that can be centuries old. Just as you and I naturally release carbon dioxide and body odors as a natural process of our daily metabolism, these corals release metabolites that send unintended cues to the sea slug larvae. Upon contact with the coral's body odor, the little veliger begins an arduous trek to the bottom of the ocean and comes in contact with the coral reef. At this point, the veliger begins an amazing transformation from small swimming voyager to a benthic slow moving lifestyle of the adult slug. This finalizes the life cycle from benthic crawler to free swimming seeker back to benthic crawler. Having chosen a good healthy reef, this sea slug has a never ending buffet for the rest of its life.

5.4 The Other Side of the Tracks

Living in a college campus community, there seem to be two types of towns rather than a singular Bowling Green. There is the student section of town (the east side) that is dominated by rental properties and apartment buildings. The night life is varied and abundant, and centers on a party atmosphere. Thursday and Friday nights are the nights for revelry as numerous front yard gatherings with loud music seem to be the common practice. The state of lawn care and car parking are chaotic at best although interspersed among the student housing are some more permanent dwellers. The other side of town (the western side) contains the downtown area as well as the more long-term residents of town. Definitely quieter at night, this part of town is home to faculty and staff that are an integral part of the University. Dividing these two areas is a train track traveling in a north/south direction such that the town is literally and figuratively sliced into two distinctive cultures. The old idiom of

"The other side of the tracks" aptly applies here depending upon the lifestyle one were to favor. Those people that prefer quieter nights and permanent neighbors favor the western part of town and those that love the night life and social gatherings are typically found on the east side of Bowling Green.

Living on the wrong side of the tracks (wrong in relation to your preference for lifestyle) could lead to a frustrating community life. We bought our house in part because we love the look, but also in part because we wanted the peace and quiet that comes from living on this side of the railroad tracks. We know our neighbors and we know what the weekends will bring. The sea slugs discussed earlier make similar choices of their neighborhood. Choosing the correct coral reef to settle on means an abundant food supply. The wrong choice for the sea slugs during their settlement phase would result in a lack of high quality food and an untimely death. The right side of the tracks, the right home, and the right neighborhood are all critical choices to make when searching for a place to live.

Of course, the term "right" is quite subjective for humans. The variety of wants and needs for our housing choices gives rise to the various neighborhoods around the world. One way to choose the right place to live is by avoiding the wrong place to live. While searching for our current home, we visited a couple of places located in the student section of town. Although they were beautiful homes, the thought of living in that area with two young children was not appealing to us. Our choice was strictly based on the visual and auditory stimuli around us. Sometimes, the wrong house can be detected by the overall odorous ambience of the house.

A colleague of mine (who reappears in more detail at the end of this chapter) relayed a story about how her father and grandfather used their sense of smell to determine the health of its inhabitants. Her grandfather was a country doctor just past the turn of the twentieth century in rural Wisconsin. The only doctor for 25 miles he spent most of his time traveling to local and distant patients to visit them in their homes. Living in northern Wisconsin, this often meant travel by horse and sled over the drifted country roads in the dead of winter. During this time, my colleague's father (who would later become a doctor himself) would accompany his father on these visits as a young boy and teen. There was no hospital in these areas, and the house calls were a necessary part of being a country doctor. Although treatment in the house was critical, my colleague's father would claim that the smell of the house was both the initial mechanism of diagnosis and the most important element for determining treatment of the patient within. The odors produced by the various illnesses would provide critical insight into the health of the patient. Unique smells would be tied to different types of diseases. Later in life, my colleague's father developed ALS, but was still so adept at diagnosing patient's illness through his nose that the Veteran's Hospital kept him on to help diagnose patients with strange illnesses.

Unlike the country environment of northern Wisconsin, coral reefs are the marine equivalent of downtown New York City. Reefs are an oasis habitat containing a wide diversity of organisms usually surrounded by marine deserts (at least from a productivity point of view). Similar to the sea slug mentioned earlier in this chapter, any larva organism being transported from reef to reef by the waves and currents in the area has a difficult decision to make. These larval organisms have to find a suitable habitat and settle on the right side of the tracks which means a healthy and productive reef system. Unfortunately for us and those larvae, coral reef health has been on the decline for decades. Due to over fishing, coral bleaching disease, and destructive sampling from the aquarium industry, these habitats have been in decline. Most notably, the overfishing has removed a large group of reef fish that function as herbivores.

Healthy reefs are free from algae and seaweed. These photosynthetic organisms (algae are not plants) tend to grow all over the corals, essentially suffocating them if they remain unchecked. The herbivorous fish acts as the gardeners of the reefs and keep these reef killers in check through constant grazing. So a reef that is devoid of these gardeners dies off because of the overgrowth. Reefs are dynamic habitats and are in constant states of growth and decline. For juvenile reef fish and larval corals, selecting a reef in decline is choosing the wrong side of the tracks to live.

Just like my colleague's father, organisms can use their sense of smell to determine the health of a potential home. Dr. Mark Hay and his students at Georgia Tech University have demonstrated that larval coral and juvenile reef fish can use their nose just like a country doctor and determine whether a reef is healthy enough to be a good home. Chemicals being released by unhealthy reefs drive organisms away and by sensing these chemicals, animals can make effective decisions about their future home and neighborhood. In the experiment, the researchers presented a smell test to over 15 species of reef fish and 3 species of corals and measured their responses. To do this, they placed the animals in the downstream end of a y-maze and then alternated the type of water flowing from the arms of the y down to the middle where the animals were. A y-maze is a fairly standard behavioral instrument for testing olfactory abilities. The maze is the equivalent of the marketing taste test, i.e., Pepsi versus Coke. The consumer (the fish in this case) is presented with two different odor types—one in each arm. By swimming away from one arm or swimming into an arm, the researcher can record preferences or avoidance (Fig. 5.2).

Fig. 5.2 Y-maze animal choice test

In this experiment, the animals were presented with either clean seawater, seawater collected from healthy reefs, or seawater collected from reefs in swift decline. The dead and declining reefs were dominated by the presence of large amounts of seaweed and dead coral skeletons. Across all species tested, the animals avoided the water from the declining coral reefs and preferred the water from healthy reefs. In a set of carefully designed experiments, they showed both avoidance of declining reef water and preference for water from healthy reefs. Just the cue from healthy coral polyps (the small little coral that is connected to all the other corals through a calcium skeleton that makes the reef) was enough to attract the reef fish. To further piece out the mechanisms involved, Dr. Hay and his students tested the preference of both coral and fish to water soaked with a common seaweed. This essence of algae (this seaweed is a macroalgae) was enough to reduce the attractiveness of the water to both species. The coral and reef fish are essentially functioning like my colleague's father in making coral diagnosis with their noses. If they don't like the smell of the neighborhood, they just move on hopefully to find just the right aroma that will literally signal home.

5.5 Nature's Halicarnassus

One of the first things I do when I get home after a long day at work is to take my dog Cedric out to the backyard of my home. The yard is a decent size for being in a small town, approximately 100×100 feet in size. I have two large black walnut trees that provide some measure of shade as well as some sense of danger. The walnut trees occasionally and randomly drop their hard nuts throughout the day, but from my perspective, I feel as if the trees wait until Cedric and I are playing in the backyard before releasing their walnuts. These nuts are about half the size of a baseball and are of similar hardness. As Cedric and I play fetch, we'll hear this thud every so often as a nut comes crashing down from the top of the tree. In addition to the trees, we have a wooden fence that encircles the yard. I am not sure the original reason for this barrier, as the previous owners placed the fence around the backyard. Perhaps for privacy or safety, the fence is now an essential element of the house for us. Shiba inus are notorious runners. So, without the fence Cedric would easily take off chasing butterflies, rabbits, and squirrels answering his instinctual needs.

Used for a multitude of purposes, fences are essentially designed to separate and contain. In our instance, our wooden fence contains Cedric within the comfort and safety zone that is our backyard. Although a nice fence, I am not really sure that the fence provides us with any measure of safety or protection. The fence does manage to keep other dogs out of our yard, but the rabbits and other smaller mammals have no problem crossing the barrier. Despite the presence for all animals, the barrier is really only effective against specific animals.

While I was in Philadelphia as a postdoctoral researcher at the Monell Chemical Senses Center, two of my favorite researchers were working on their version of a chemical barrier or fence. Drs. Larry Clark and Russ Mason were technically employed by the National Biological Survey at the time although they did their research at the Monell Center. Their task was to create an easy and cheap chemical barrier to keep birds out of landfills. Within our modern culture, there are places we like to have available to us, but just not near where we live. Two such places, landfills and airports, are essential elements of modern society, but are elements that nobody really wants to live next to. Airports are unappealing because of the noise and landfills because of the smell. Unfortunately, city planners have a habit of placing landfills and airports near each other and as birds visit the landfill for free meals, there is an increased possibility of these birds getting sucked into jet engines with dire consequences for both the bird and jet.

Drs. Mason and Clark worked to develop a chemical repellent that could be added to landfills that would be the chemical equivalent of my backyard fence. The repellent would allow some organisms in, but would keep out the birds that could potentially cause air traffic accidents. They found that the chemical capsaicin, familiar to most of us, worked pretty well. Capsaicin, also the active ingredient in pepper spray, is the compound in chili peppers that gives the hot and burning sensation to food when used in cooking. Buffalo wings, spicy Chinese food, and good Texas chili all use capsaicin to give the dish that extra kick necessary to delight (and punish our mouths). Capsaicin activates the trigeminal sense (mentioned in Chap. 2) and provides just the right level of irritation that birds tended to avoid visiting areas sprayed with the irritant. As a side note, this irritant also works for those pesky herbivores that tend to munch tender plants in gardens. Thus, the capsaicin forms a chemical fence around landfills and backyard gardens. The only downside of the capsaicin barrier is that the compound needs to be touched or consumed before the irritation actively works on its target. Other animals have developed a grisly olfactory fence that protects their most precious assets.

In the first book (A Game of Thrones) of George R.R. Martin's bestselling series "Song of Ice and Fire," the audience is led to believe that the main character of first book, Ned Stark, will be the hero of the series. True to Martin's writing style, this character is summoned to the main palace and summarily beheaded for treason. As is the common practice for this fantasy time period, Stark's head is placed on a pike and mounted near the entrance of the castle. Here, the head will serve as a symbol of danger to those that want to follow a similar treasonous path. Provided there are enough heads around the palace, this fence of treason is a strong visual signal not to enter the premise with thoughts of overthrowing the King. The fence serves to (hopefully) protect the precious life of the King.

A new species of wasp performs a similar trick to construct an effective olfactory barrier for its larvae. The "bone house" wasp behavior and nest were recently discovered by ecologist Michael Staab in southeast China. The bone house wasp is within the family of spider wasps. Spider wasps have very powerful stings and use these stings to hunt down their deadly prey. A truly wondrous group of insects that include members such as the Tarantula hawk, these spiders usually sting their spider prey and drag the bodies back to their nest of eggs. In a gruesome act that would make even Martin's "Red Wedding" look tame, the female eats all of the nonessential parts of its victim, leaving the spider alive but immobile. The female then places the spider in its nest in order to feed the hatching eggs, providing a delicious first meal for the juvenile spider wasps.

The bone house wasps seem to have taken this macabre act just a little farther than its fellow spider wasps. This wasp either collects or hunts predatory ants and summarily kills the ants. Wasps, as a general group, have a variety of mechanisms which have evolved to protect or hide their eggs. Those wasps that dig burrows or holes often have multiple chambers to the burrow. There may be multiple chambers that hold eggs called brood chambers, but independent of the inner design, there is a vestibule or outer chamber without eggs. Different species of the wasps will fill the vestibule with twigs, dirt, or leaves in order to visually hide the nest from potential predators. The bone house wasp has a similar house design with egg chambers beneath an outer vestibule. This wasp drags those dead ant bodies to its nest and fills the vestibule with the corpses of the dead ants. The bone house wasps build its own mini-mausoleum: the wasp version of the famous wonder of the ancient world, the mausoleum at Halicarnassus. Their Halicarnassus produces chemical signals from the dead and decaying ant bodies that serves to warn other predatory ants to stay away from nest because this is a site of ant death. In addition to warning predators, the nests of the bone house wasps are attacked less by parasites. Dr. Staab believes that chemicals emanating from the decaying ant bodies serve a protective role against parasites also. As pointed out in previous chapters of this book, ants produce a wide variety of chemical signals (through cuticular hydrocarbons) that are used for nestmate recognition, slave raids, and species recognition as well as other sorts of messaging within ant communities. These hydrocarbons, because of their chemical properties, tend to stay attached to the ant bodies and as such are very long lasting chemical signals for the wasp nest. In the first study on these wasps, the authors noted that the most dominate type of ant corpse found within the nest was a particularly large and aggressive ant species found locally. Maybe the bone wasps are specifically selecting the most aggressive and feared ant species and using their corpses as a chemical signal to other predators. If true, then the bone wasp is nature's equivalent of "staking the head" (as in *A Game of Thrones*) of the most dangerous usurper outside of their nest.

5.6 The Book of Odor Memories

I was recently writing sections of this book in a campus coffee shop when one of my colleagues stopped in to get coffee. She spotted me holed up in a corner and sat down at the table next to mine. I had spilled out all of the contents of my bag across the table top. Among the mess was my computer open to a different chapter of this book, a number of papers scattered about for quick reference, and a large black ice tea. I believe that she had come by to do some reading and grading of her own. She is a faculty member in the College of Music here on campus, and we have known each other for several years if not decades. We were at similar points in our careers in that we had been successful enough as faculty members to be tapped to serve on several

University-wide committees. Serving on committees at a university is both a blessing and a curse. At times, there are chances to make significant changes to how we educate our students, yet at other times, working at the juncture of fellow faculty members, student's needs, administrative desires, and state guidelines is a quagmire that can threaten to darken even the most optimistic person. My music colleague (Mary) always seemed to approach these appointments with an excellent mixture of intelligence, pragmatism, humility, and insight.

Early on in my career, I remained figuratively locked in my biology building working on getting my scientific laboratory up, running, and productive. I rarely ventured out of the building, let alone actually walked over to other departments to have significant discussions with faculty in other disciplines. Luckily for me, I took a part time administrative position, after tenure, which forced me out unto the campus and into discussions with the different faculty across campus. I was fortunate enough to run into this colleague and have had numerous enlightening conversations on personal life, teaching, and most importantly, music. Music of all types is a joy to me and having in depth discussions on music, writing music, and playing helps further this passion.

As she sat down at her table next to mine, she asked me what I was working on. I described the outline of this book and the audience to which it is intended. In addition, I quickly outlined a couple of animal stories on olfaction and taste to give her just a flavor of what I was attempting to do. As I recanted some of the biology that I had already written down, her eyes began to light up. After I finished my cursory overview, Mary provided me yet another story of home that I had to capture here.

As I mentioned above, her father and grandfather were medical doctors and her father was in the Navy. As with most military individuals and in particular, those in the Navy, travel is a common aspect of their service. During the Korean War, her father was stationed in Japan working at a military hospital. This was before marriage and kids, so most likely a time in one's life where exploration could be a key element of life. I, myself, am intensely interested in Asian philosophy, art, and literature. So, being stationed in Asia with free time on my hands would be a particularly interesting time. Apparently, during his time in Japan, Mary's father fell in love with the local art and began to collect several items that would serve as a reminder and memory of his time in Japan. A formative time in his life, he returned to the states with several books on Asian art.

Fast forwarding several years, Mary is now a little girl and living with her parents in an apartment. In one corner of the apartment, two book shelves lined the two side walls which made a perfect corner for someone interested in a quiet area to study or play. In addition to the books and in an apartment where space was a premium, a grand piano was pushed into the corner. Apparently hiding under the piano next to the book shelves was the perfect place for my colleague to sit and play with her dolls or take a nap. Being too young to read at the time didn't stop her from pulling books off of the shelves and looking at the pictures. One book that she kept returning to was an old Asian art book. The pictures of the water color paintings and sculptures were beautiful to study even for such a young girl. Even the writing, Japanese Kanji, was artistic in style and held her interest. Maybe this is why she became an artist. Spending days underneath a grand piano, perhaps listening to the occasional piano piece, and "reading" books on Asian art had to have some lasting influence. Apart from these visual and auditory stimuli, she was drawn to the intriguing smell of the art book. The book presented a new and exotic smell to her, something she had not smelled before. Unlike the singular voice of the piano, the smell was a mixture of odors that intrigued her. A little sweet, a touch of musty, and some other aromas emanating from the pages held her interests. Imagine being a child, small enough to have your own home under the piano, reading and smelling to exotic far away smells, and images with the occasional serenade form a grand piano. She felt so grown up and yet so safe in her special home. A lasting childhood odor memory that reminds her of home.

Years later, after her father passed away, her mother sold their old home and some of the belongings in it. My colleague made sure to take that old Asian art book. She still has that book now. The faint odors still remind her of that time in her life, her father, and her special "home" underneath the piano. The fragrance quickly takes her back to that time and to that place and evokes memories that flood her mind. This is the power of odors and the power of memories of home. The smell of loved ones, of comfort, and of safety.

5.7 The Smell of Home

Although there is not an exact scientific study that demonstrates this in humans, I would imagine that our blood pressure and stress level drops upon sensing the smells of home. Science has found odors that raise blood pressures, which are usually obnoxious smells, and odors that indeed lower our stress levels. Some of the fragrances that succeed in altering our stress include mango, lemon, and some floral odors. Perhaps the impact of these odors on our physiology is why these odors are common within household products. We use cleaners, candles, and sprays to either cover up unwanted odors or create a certain ambience of home. As we grow up, we learn to associate these odors with the comfort of home similar to my colleague, her piano sanctuary, and the Japanese art book. Our home is a unique combination of people, pet, food, and plant odors that we have learned to associate with emotional states. Studies have shown that odors collected from homes (random homes to those smelling the odors) did lower the depressive moods in the participants in the study.

In the beginning of the chapter, I described the odor sensation that greeted my nose as I walked into my home after a day at work. I can tell whether my wife has beat me home because I would smell the beginnings of supper. If I get home first, then I start the cooking process and she is greeted with the aroma of dinner. Sometimes candles are lit in order to create a different odor environment. If I am writing at home during the evening, then there is a vanilla candle in the corner that I light before writing. It seems as if the aroma is the pièce de résistance for my writing nook in the living room. Whether the stimulus is an Asian art book, lavender air freshener, or a vase of roses, we can create an environment that signals home to us.

Certainly those companies that produce products allowing us to create that environment are fully aware of our need to create that signature environment. In the 1950s and 1960s, the clean and safe home smelled of ammonia and bleach. During this period, our homes were refuges away from the potential germs and virus that were ever threatening from the outside world. Counter tops weren't safe unless they smelled of bleach. As a child of the end of this era, I remember thinking that public bathrooms were safe to use if I detected either ammonia or bleach from the newly cleaned floors. As the 1960s gave way to the 1970s, the quintessential home odor changed. This change in odor preferences closely followed the ability of commercial chemists to produce different odors in products. Pine and lemon scented products dominated the cleaning and fragrance market. Supposedly, the pine reminded us about the great outdoors and that we were quite advanced in our technology by bringing that outdoor smell inside of the house. Lemony fresh was a marketing scheme used by all sorts of products. Obviously, citrus and pine were signals of home or at least, that is what we became conditioned to expect or believe. Over the last three decades, fragrance chemists have found either the scents of natural items or recreated odor mimics that fool our senses. Floral scents have erupted with this field and allow home owners to forgo the actual flowers to use a spray instead. At this point in time, these products have gone a little overboard with the creation of odor melodies. In one of my recent trips to a store with candles, I found a Thanksgiving Day candle which was a combination of pumpkin pie, cranberries, and fall leaves. Some other odors, including a spring rain, the Amazon rainforest, and ocean breeze, are probably some random combination of chemicals put together and when smelled reminded someone of some day at the beach or after a rain.

The choice of these odors and their association with our homes constructs an emotional atmosphere. As pleasant odors increase, our level of contentment and happiness also increase. If the unpleasant odors dominate our house, say when a certain black lab tussled with another skunk, we can get tense. In essence, there exists the possibility of subtly influencing the emotional environment in our homes by altering the odors and fragrances we use. We choose these odors based on the emotional reaction that they evoke although we aren't fully aware of this reaction. Pick the right candle and everyone is awake ready to greet the day with vigor. Another air spray at night for a quiet book discussion around a fireplace. Home is more than a building or shelter, it is a collection of aromas.

So important is this connection between odors and home that a recent movement among realtors is creating personal fragrances for the homes that they are selling. Among the high end real estate market, custom odors for signature homes are being used to connect potential buyers with a home. A new company in New York now designs suites of fragrances that are designed to evoke certain feelings. For condos in Miami, the scents of ocean breezes are piped through the lobby even though the condos are not near the ocean. Special machines can be placed throughout the common areas to project just the right feeling. These tricks are a high tech and updated version of placing a freshly baked loaf of bread in the oven or plates of warm cookies on the dining room table that realtors have used for years. These "older" odors were meant to envelop buyers with a feeling of warmth and comfort. Now, odors such as floral scents for common areas and earth tones for workout rooms are common place. Thus, each room has a unique emotional feel that matches the purpose of that room furthering that connection between person and place. This practice has even extended beyond homes into other locations. Calming and soothing odors are used in hospitals as well as artificial outdoor (woody and cinnamon odors) scents piped into amusement parks that serve both to cover up the prevailing offensive odors and to help place customers into the right frame of mind. Good homes are a place of emotional peace and that emotional peace begins at the tip of our nose.

Chapter 6
The Cocktail Party of Life

The Chemical Conch Shell

During my morning writing rituals at the bakery where my creativity ebbs and flows, I often find myself totally engrossed in my craft such that the only thing I am consciously aware of outside of my writing is my pumpkin muffin and occasionally the background music. Then there are those moments that my muses leave me and my writing comes to a crashing standstill. During these times, I turn to observe other customers who frequent the bakery each morning. By allowing my mind to wander from the writing at hand, I become captivated by the social interactions of those around me.

As a consequence of my scientific curiosity, I find it fascinating to observe individuals interacting with one another. One of the more interesting aspects of observing humans is social behavior, particularly the subtle, nonverbal ways in which people communicate. I can notice changes in body postures, slight differences in the tone of voices, and even changes in facial expressions. Apart from the actual words spoken during these interactions, these other nonverbal forms of communication can provide a wealth of information about the relationship between the people speaking, their emotions, and maybe hidden meanings behind the words they are using.

Coffee houses and bakery really serve two purposes; the most obvious one is to sell drinks and pastries to waiting customers. Secondary is the atmosphere and community that is built allowing the majority of people at the bakery, the ability to socialize with others. Quite often several individuals from the University gather for a morning meeting over coffee and bagels. A corner of the bakery is occasionally inhabited by a small group of older women that always seem to be having an enjoyable morning. Other people spread papers out covering the entire surface of the table. Once the papers are in place and the coffee is purchased, a conversation ensues that appears to be of a great deal of importance to the participants.

True to my scientific curiosity, I catch bits and pieces of their discussion and study their interactions. I try to reconstruct conversations by studying the subtle body postures and hand gestures of the speakers. I study their facial expressions and where their eyes are gravitating. The tone of their voice is also a continual clue that allows me the opportunity to attempt to piece together the social meaning of these interactions.

P.A. Moore, *The Hidden Power of Smell*, DOI 10.1007/978-3-319-15651-4_6

The real meanings or purposes of these meetings are, of course, beyond my knowledge. I know that, quite independent of the intended purpose of the gathering, there are social interactions occurring at a level that is deeper than just the words that are spoken. At one meeting, a tall gentleman dressed in a dark blue business suit with a red tie is clearly dominating the proceedings. Even from a distance, the person in charge of this meeting can be gleaned from the appearance of his power suit and tie. Although, in terms of physical size he is no different from others, he is establishing the forcefulness of his points by increasing the intensity of his voice. Everyone's eyes are glued to him, and his face is rigid with deep seriousness. The other participants at his table appear to submit to his ideas.

At yet another table, two casually dressed individuals are discussing some papers. As the conversation goes back and forth between them, they shift in their seats and use their hands to stress points or to gesture toward some of the papers. A stark contrast from the previous scenario, a pleasant and constructive conversation is occurring where they appear to give and take equally in the discussion. Every once in a while, a small smile creeps over one of them or a slight laughter emerges.

Over in another corner sits a couple that is enjoying a slow and relaxing morning. Their body postures and facial expressions show that a completely different interaction is occurring here, in sharp contrast to the previous tables. The gentleman reaches across and lightly touches the woman's hand; there is a deep look in her eyes as she glances at him. They both lean into the table in an effort to be nearer to one another.

Each of these social interactions has a very specific purpose. Although I am not privy to the actual conversations during these meetings, I can deduce a little meaning from their interactions through careful observation. The actual words spoken don't matter all that much to me, yet the nonverbal communication and behaviors observed during these interactions are intriguing. If I was not so immersed in biological thought, I would probably be a sociologist in order to study group dynamics and the origin of social behavior in humans. However, since I study some aspects of social behavior in my research life, I can combine both my love of biology and my interest in social interactions. Besides which, I get far fewer stares from people when I closely watch animals interact as opposed to my inquisitive periods in the bakery. Although we are very social, we are hardly unique with respect to the rest of nature in regards to social behavior.

Many various animals exhibit complex social behaviors and, as a consequence, form elaborate animal societies. Probably some of the most well-known social structures occur within mammals. Many primates live within family units or even extended groups. Most marine mammals have some form of social organization, usually forming pods, and even smaller mammals, like mice and rats, have a system of social interactions that form a society. What may not be known is that social behavior is also very prominent outside of mammals. Fish often school or travel in large groups that require some aspect of social behavior, at very least, in coordinated movement. Lobsters and crayfish form hierarchies and that are reinforced with aggressive behavior that dwarfs the typical head-butts seen in mountain goats. What separates human social behavior from other forms found in nature is the unique ability for humans to have multiple social systems with differing hierarchies. Some of my days at the University are a case in point.

6.1 Social Hierarchies

It is a Friday morning as I arrive to the lab and what ensues is a day of meetings. The course of meetings will take me through the different aspects of social interactions with my status fluctuating from being the top dog to runt of the litter in a matter of hours. I start the morning with my weekly lab meeting which a large group of students who themselves are at various stages of their academic career. I currently have 15 people doing some aspect of research under my guidance, which creates an interesting set of social hierarchies. There is a senior research associate that has just joined my lab. Although technically "underneath" me in regards to the administration of the lab, he is my senior in both age and scientific knowledge. Next in line are the graduate students, including a mixture of Ph.D. and Masters students, which I advise. Finally, there are a number of undergraduates some of which have 3 years' experience working with me while others have been around for 2 weeks. Although I do my best to treat everyone as equals, there is always a series of social interactions that ultimately result in the formation of a hierarchy. Since I run the lab, I am ultimately at the top and all of the final decisions rest with me. My place within this hierarchy is my social status and as we shall see, this status is highly dependent upon the group with which I am surrounded.

Next on my calendar is our periodic faculty meeting. Anyone who really wants to see social dynamics should be a fly on the wall during a faculty meeting in a university department. Our faculty consists of four different levels of appointment. The assistant professors are the youngest appointees and have been in the department less than 7 years. Next on the ladder are the associate professors and the top rung is reserved for full professors. Above us all is the head of the department, our chair. The chair position carries the final authority for all departmental decisions and one could argue that this duty makes the chair an alpha position within our department. Since I am a mid-to-late-career professor at the moment, I leave my lab meeting as the top dog and walk into the faculty meeting where I reside somewhere near the middle on a power rank.

My final meeting of the day is that of a University-wide committee that is made up of undergraduate and graduate students, professors, and high ranking University administrators. In some ways, since the faculty do not pay tuition to the University and ultimately serve as financial drains on the system, faculty members are on the bottom rung with this group. Being the only non-administrator/faculty member on the committee, I have finally sunk to my lowest social standing of the day.

The development and the maintenance of these many hierarchies is one of the most important aspects of studying social behavior in humans and other animals. In fact, one could almost say that without hierarchies there would be no social behavior. Whether that being is the queen in insect societies or the queen in human royalty, the main concepts underlying social interactions, social standing, and status, are the same. Any level in the hierarchy carries with that position certain privileges or advantages that other positions do not. The higher up a position is within the hierarchy, the better these privileges or advantages become. Being on top of the hierarchy means more mates, better shelters or territories, or more food. For example, the queen bee or queen ant is the only individual that reproduces for the colony or hive. All of the workers and guards are there to serve the needs of the queen, whether they are supplying the queen food or caring for her eggs. A frequently used quote on the benefits of a high social status comes from the movie, *History of the World, Part I*. Mel Brooks, portraying a lecherous French monarch, repeatedly states "It is good to be da King" as he orders around servants and wenches. Just as the "top dog" gets all of the privileges in animal societies, one need not look too far to see the benefits of being higher-up in human hierarchies, whether they are business or royalty hierarchies.

6.2 Do the Clothes Make the Man or Does the Man Make the Clothes?

Growing up in a middle class household, my parents often impressed upon me that the clothes make the man. I would hear this idea in many different forms ranging from "dressing for success" to "you can tell the cut of his jib by the clothes that he wears," which can be understood by the nautically inclined. All of these statements reflect the sentiment that one can gather some meaningful information about the social standing of an individual simply by how he or she chooses to dress. Underlying this thought was the idea that those individuals that were to be respected within society, the doctors, lawyers, politicians, and business individuals, would surely choose to dress in professional clothes.

Evolving from these ideas were the business principles behind the idea of power suits and ties. The color red would send one message, whereas the color yellow would send a different message. When attending that all important business meeting, the red tie with dark blue business suit would exude confidence and power. Instant social standing is assured just by having the right combination of colors and clothing. This raises a subtle but interesting point: Do the clothes make the man or does the man make the clothes? Or in social (and non-sexist) terms, do the clothes reflect the social status of the wearer or do they determine the social status of the wearer?

As humans, we often want to think of ourselves as a highly evolved species that would certainly see beyond the simple coverings of clothes and be able to judge the true nature of an individual regardless of what they are wearing or how they smell to us. Subscribing to the contradictory, yet equally pervasive adage "Don't judge a book by its cover," we would think that putting a crown on a vagabond does not make her or him a queen or king. This person would still be a vagabond. This social standing has to be earned or inherited. Certainly the abilities of an individual are a manifestation of who they are or what training they have had and have little to do with their appearance. Thus, the clothes merely reflect the status of the individual, right?

If this were true, then certainly if we went to a new doctor, our level of comfort or trust in their abilities would not matter if they were wearing nice clothes or ripped jeans and a ratty old t-shirt. Imagine walking into a business meeting where the

CEO had not showered in days or wore clothes that we would call "inappropriate" for such an important meeting. Our perceptions about people, especially those initial perceptions, are colored by the signals that clothes and even perfume or cologne send to us. But how big of a role do these nonverbal signals play in our impressions of individuals? Are these the conch shells of The *Lord of the Flies*, where whoever has the shell is currently the leader and, thus, the shell determines the social status rather than reflecting it? In reality, the answer is whoever has the right chemical perfume rules the roost.

Allen Moore, at the University of Georgia, has discovered nature's version of *The Lord of the Flies'* conch shell by studying the development and maintenance of social hierarchies in cockroaches. Apart from our human bias concerning our perception of the cockroach, they are exquisite research animals. This is especially true if you disdain the ordinary American variety of cockroach and chose to study the forest cockroaches of Tanzania. These cockroaches engage in very intense social interactions, which include fighting and ritualized courtship. A fight (termed an agonistic encounter by scientists) between cockroaches hardly follows the gentlemanly rules of boxing (called the Marquess of Queensbury rules). The agonistic encounters include butting, lunging, biting, kicking, grappling, and chasing. Simply putting two cockroaches together in an arena will result in all-out war. Within seconds, one animal will ram, bite, and kick the other animal. All of this occurs rather quickly and ends with one animal, the dominant, chasing the other, called the subordinate, around the tank. This hierarchy, and other more complex ones that form among groups with multiple animals, can have profound consequences for the mating behavior of the cockroach. Often, a specific animal in the hierarchy (and probably not the obvious one to the human observer) is the most attractive to the female cockroaches. Even for the lowly cockroach, reproductive and social success is tied to the proper place within your society. But before we can reveal the secret of the female's choice, a little more background information about cockroach agonistic interactions and insect behavior is needed.

Anyone who has watched a nature show will eventually come across the typical video footage of intense social interactions. These often take the form of the most violent, making spectacular scenes. These images tend to be the loud and powerful head-butts of mountain goats, the impressive antler crashing of male deer, or the loud roars and sometimes lethal consequences of elephant seal fights. One common feature that is seen in the fights from this diverse group of animals is that fights are relatively long. Determining the alpha male or dominant animal can take an incredible amount of time and energy. Yet, Dr. Moore realized that his cockroaches fought intensely, but only for a relatively short period of time. In fact, he began to think that these fights were not the typical fights seen in other animals. The fights look as if the outcome was predetermined before the cockroaches began to fight, similar to a match between a heavy weight boxing champion fighting the proverbial 98-pound weakling. Yet the cockroaches in Dr. Moore's studies were always the same size. This begs the question as to which cockroach is the 98-pound weakling and how the contestants could know the difference. To answer this question, the role of chemical communication in insects needs to be reviewed.

One of the most prominent themes in insect biology is their reliance on chemical signals for making a host of ecological decisions. Insects have been shown to use pheromones for mating, individual recognition for social hierarchies, to control reproduction, and a variety of other behaviors essential to survival, such as the nest mate recognition exploited by the thieving ants described in Chap. 1. In fact, it would be possible to center every nature story in this book on insects. Cockroaches, just like any other insect, use chemical signals in their daily lives and have prominent antennae on their head that are sensitive to a host of different chemicals. Thus, Dr. Moore thought that, instead of a series of vocalizations used in many different mammalian fights, the cockroaches may be using a language of smell to communicate who the winner and loser will be in the fights.

It turns out that these Tanzanian cockroaches produce a pheromone that communicates clearly to other cockroaches who is the boxing champion and who is the 98-pound weakling. As with most other insect pheromones, the signal is not a single omnipotent compound, but a mixture of three different chemicals that is analogous to a sentence constructed of various words conveying specific information such as social status. Each of three different components conveys a different meaning for the cockroaches. Moore took each of the chemicals in the pheromone blend and placed them on tiny pieces of paper (almost like the clothes analogy above). He placed these pieces of pheromone paper on cockroaches as either a single chemical, or in the many combinations of two, or all three components. The mere presence of one of the chemicals makes a particular male cockroach act like a subordinate: that cockroach cowers and runs for cover. If one of the other two chemicals is placed on the same cockroach, the opposite happens. The cockroach now struts around the arena as if it was king of the world. If both of the latter two compounds are attached to the cockroach, the effect is magnified. This signal is an essential component of these social interactions. Cockroaches produce all three components of the signal; therefore, the presence of the signal does not make one cockroach dominant and the other one subordinate. The relative concentrations of the three components is the critical aspect that makes the difference in the individual's social standing.

In the English language, sentences are constructed from three simple building blocks: the subject, the verb, and the object. Basically, these three concepts are: who is doing something (subject), what they are doing (verb), and to whom are they directing the action (object). The meaning of some sentences can change dramatically by emphasizing one of these components over the other two. For example, let us examine a simple phrase like "I am King." If the "I" is emphasized, we are trying to differentiate between the fact that "I" am the ruler as opposed to any other possible usurpers. If the "am" is accented, the sentence takes on the connotation of elation as if reveling in achieving the highest stature. Finally, by emphasizing the "King" and raising our voice a little, the sentence may become a statement of surprise or even a question. This last example certainly carries less forcefulness than the previous two versions of the sentence. In our running analogy, the three words are different articles of clothing with the word "King" being the powersuit.

Like the analogy "I am King," the cockroach pheromone system acts in a similar manner. By emphasizing (having a higher concentration of) the "I" and "am" portions of the pheromone, the cockroach strongly convinces his opponents that he is indeed the king and that they should cower at his 6 feet. Conversely, by accenting the "King" (having a higher concentration of this component), the cockroach is asking the question "I am King?" and is thus relegated to a submissive role within the population. For the cockroaches, the relative concentration of these components makes an animal dominant, and these signals are just as clear as grabbing the conch shell as in *Lord of the Flies* or putting on the one ring in *Lord of the Rings*.

If being on the top of the social ladder confers advantages in regard to mates, shelter, and food, why would the cockroaches produce the third component that relays a message of submission? Why not just produce the two dominant components of the pheromone and become super dominant? This approach should get the male cockroach all of the matings he could ever want, right?

Just as there are stereotyped agonistic interactions between males, there is also aggression between males and females during the courtship that precedes mating. The super dominant males will win all of the fights, but the problem is that they get too aggressive for the females. During fights between males or between males and females, one of the animals may lose one or both of their antennae. For humans losing our sense of smell is not as critical to our survival as losing our sight or hearing, but for the cockroach, the loss of an antenna is the loss of their whole sensory world. Without the sense of smell, they cannot mate, locate food, or find places to lay eggs. Thus, females choose males that are dominant, not super dominant so as not to get injured during the intense courtship ritual. The third pheromone component serves to temper the social effects of the first two pheromone components. Just as the clothes make the man, in the example above, the smell makes the cockroach. Without this perfume, the cockroach is nothing, but with it, the cockroach is king.

6.3 The Secret (Smell) of My Success

The cockroach use of a pheromone in determining the status of the individual is somewhat unique within nature. The more common situation is when an animal becomes dominant and consequently the dominant animal produces a pheromone that reflects that status. We can think of this situation as a perfume that reflects the social history of the animal. Dominant animals would wear Chanel No. 5 whereas subordinate animals would maybe choose a knock off brand bought in a gas station. There is a good reason the preceding sentence was phrased as a conscious choice on the part of the animal. It is as if one could choose to either wear the upscale perfume as opposed to the garden variety smell. If this is the case, why would any animal choose the subordinate smell and, as a consequence, gain access to only the sub-par habitats, food, and mates?

Crayfish, as described in the previous chapter, are one of the many crustacean models for animal behavior and chemical communication. They are nocturnal animals often found in murky or muddy environments. In these environments, particularly at night, visual signals are of no consequence. Crayfish are also gregarious, living in large groups, and often in very high densities. They have a bladder system (as described in the previous chapter) and two enlarged claws used for fighting and social behavior. All of these characteristics make them ideal for studying social behavior and chemical signals.

As with cockroaches, crayfish will often fight and establish hierarchies but, unlike the cockroaches, the agonistic interactions between crayfish can last for several minutes. These fights can put to shame the brutal nature of mountain goat headbutts or elephant seal fights. The enlarged claws are used to grasp and tear; while some of the fights between crayfish can even turn lethal, more often than not the fights usually end before this occurs. The outcome of these fights is used to establish a hierarchy within the population and, as with most animals, the larger the animal, the more dominant that animal becomes. The smaller animals tend to be at the bottom of the hierarchy and are thus relegated to less than ideal habitats and have fewer opportunities to mate. Social signals in crayfish are closely tied to the neurochemistry of the animals' brain.

The crayfish nervous system is an elegant model to study the neurochemical basis of hierarchies and dominance, which all revolves around a simple organic molecule known as serotonin. Serotonin is a very interesting molecule and is associated with a host of behavioral and neurological problems in humans, such as schizophrenia and Parkinson's disease. However, what is interesting is that serotonin is linked to social behaviors in both humans and crayfish. Here is where I would like to introduce Don Edwards, who has been leading the research into the role that serotonin plays in the social behavior of crayfish at Georgia State University.

Dr. Edwards has been examining the levels of serotonin in animals with different social histories and altering the serotonin levels in other animals in order to understand the linkage between serotonin and dominance. Dominant crayfish tend to have more serotonin in their nervous system than subordinate crayfish. To state that serotonin levels directly correlate with dominance status is not the complete story but a simplified version for the purposes of a book on chemical communication. Basically, Dr. Edwards has shown that as serotonin increases in concentration within the nervous system, the subsequent dominance status also increases. A number of factors such as past social experience, size, and sex of the individual also play a role in dominance establishment and can influence the effect serotonin has on the nervous system.

Crayfish have a semi-open circulatory system. When serotonin is produced by the nervous system, the molecule is released into its circulatory system and delivered to appropriate places within the crayfish. This is in contrast to the closed circulatory system of mammals. In a closed system, the body's blood is enclosed in a system of pipes, which we call blood vessels. The blood is forced through these vessels by the repeated contractions and expansions of the chambers of our heart. In a "true" open circulatory system, there is no system of pipes and no pump serving to push the blood around. Crayfish have a semi-open circulatory system because they have a small set of open-ended blood vessels that are connected to a rudimentary heart. This type of circulatory system is critical to the development of our story on crayfish, social signals, and serotonin.

Now that we have established a base of knowledge for the biology of crayfish and serotonin, let us return to the idea of chemical signals. If you were to place two crayfish in a fish tank, you can guarantee that they will fight. A casual glance at the fight with untrained eyes probably would not reveal anything unusual. One would see two clawed animals pushing each other around with the claws closed. If one of the crayfish does not back down from this "boxing match," the fight will begin to escalate. The crayfish will then open their claws and actively grab hold of each other in an effort to turn over and pin their opponent. Professional wrestlers have nothing on these animals as they push, grab, and use leverage to try to influence the outcome of the fight. Again, if one of the crayfish does not submit, they both take the next step in fighting: an all-out unrestrained, no-holds barred fight with the lethal claws. At this point, any fighting rules go out the window and each crayfish will literally attempt to dismember their opponent. During these increasingly aggressive stages of fighting, one of the animals usually realizes that winning the fight is not in the cards for the day and retreats from the other animal.

Notice that I used the words "untrained eye" when describing the observation of these fights. There is far more to these crayfish fights than the obvious pushing, grabbing, and tearing by the lethal claws. The trained observer will notice subtle but key elements of the fights are missing from the description above. During the fights, crayfish will often "flick" their lateral antennules. The lateral antennules are a pair of small, hairy appendages that stick out from the top of their head. The "hairiness" of the appendage is due to the thousands of chemoreceptors that are located on the appendage, which is often described as the nose of the animal. (Although a more appropriate description would be one of the noses since the crayfish carries 12 different appendages, each of which functions as a "nose"). The "flick" of the crayfish antennule is a rapid downward movement of the appendage and is the equivalent of our sniffing to gain more information about a particular smell. Thus, the crayfish are actively sniffing some chemical in the environment, perhaps assessing the smell of their opponents during their fights. But what do they smell?

Back in Chap. 1, I mentioned the story about "the smell of fear"; as you may recall, crayfish have a bladder in which they store "urine." In addition to a bladder, the crayfish have a pair of nephropores, one directly beneath each of their eyes. The nephropores are the outlets of the bladder and can be opened and closed to release their contents into the surrounding water. When opened, a stream of water is expelled directly forward and away from the animal. It may seem awfully strange and unsanitary to us to have your nephropores right beneath your eyes, but this location serves a very useful purpose to the crayfish during social interactions.

During a fight, two crayfish face each other and carefully approach with claws raised and open in a threatening gesture (Fig. 6.1). What is obscured from view by the presence and use of such lethal weapons is the exchange of a wealth of information about the opponent through the use of urine and the chemicals contained within it. Instead of solely relying on the visual information about the size and strength of their opponent, these crayfish begin a literal "pissing match" and release relatively copious quantities of urine in the direction of their opponent. With the open circulatory system, any change in serotonin will also have a concomitant change in the

Fig. 6.1 Crayfish flicking antennule

amount of serotonin and metabolites of serotonin in the blood of the winning crayfish. Eventually, this serotonin or its metabolites will find its way into the bladder system of the crayfish to be used in its next match giving the winner of the fight an added boost.

The subsequent role of these stored hormones (in crayfish or other social animals) can cause what are known as "winner" effects. This "winner effect" is the phenomenon where a crayfish that has had a recent winning fight is more likely to win the next fight. Winners keep on winning and losers keep on losing (because of a "loser effect"). This is analogous to confidence, ego, or momentum in the human world. Think of all the times that we have heard about a sports team believing that they are invincible. If two crayfish are placed in a tank and one of them has just won a fight, the winner will invariably win the next match. If this experiment is repeated, but this time the urine release of the winning crayfish is blocked, the "winner effect" disappears and the two crayfish act as if there never was a previous fight. So, the momentum or ego of the crayfish appears to be tied-up in the chemicals being released in the urine. The urine "announces" the win–loss record of crayfish just as surely as a ring announcer does in a heavyweight championship fight. But the urine has a more far-reaching consequence for the crayfish.

Imagine a situation with the classic schoolyard bully. Everyday the bully enters the schoolyard and eyes all of the potential marks. Seeing a smaller kid over by the swings, the bully approaches and begins to flex his muscles or starts to make some threats. Quickly, the smaller individual will succumb to this level of intimidation and will hand over his lunch money. If this scene is continued day in and day out, two types of phenomena will manifest themselves. First, the smaller kid will recognize the bully at a distance and will either hide or relinquish his money to the bully without the show of muscles or verbal threats. A certain relationship has been established, and this relationship needs very little reinforcement to remain in place.

Second, the smaller kid will begin to exhibit subordinate behavior all of the time, not just to the bully. The pattern of bullying will result in a lack of ego or low self-esteem in the kid being bullied.

This situation is not uncommon in nature and is just an extension of the winner or loser effect described in the previous paragraph. Another example, albeit a little unpleasant, occurs when a dog has been mistreated for a period of time by abusive owners. If one is in the presence of a dog that has been treated poorly, the dog will exhibit classical subordinate behavior to everyone (not just the previous owner). This is seen when the dog approaches a human and often keeps its body close to the ground, its tail is tucked underneath the body, and the dog often refuses to make long-term eye contact with the human. This is a series of conditioned subordinate behaviors and usually occurs with long-term exposure to an overpowering physical presence, such as a schoolyard bully or abusive owner. This behavior is also seen in crayfish. What is unique about crayfish and our story on urine is the ability of only the smell of urine to produce the same low self-esteem effect seen in dogs and bullied school children.

Here, with great pleasure, I can count on my own work. This work was performed with Dr. Dan Bergman, a former graduate student of mine. If you subject a crayfish to a week-long series of repeated fights with larger crayfish, the original crayfish will perceive itself a loser and will begin acting like our dog or bullied child described above. The "loser" crayfish will lose fights, run away from opponents, and generally act as a thoroughly subordinate crayfish. This seems to be a fairly straightforward concept. Now, if you subject a naïve crayfish (naïve means no social history) to a week-long exposure of only the urine from a bully crayfish, the naïve crayfish will become a subordinate crayfish and act identical to the crayfish that has been beaten up for a week. Just the mere smell of a bully crayfish is enough of a stimulus to alter the behavior of the second crayfish. This phenomenon exhibits itself without any physical interaction and without any visual contact with the bully crayfish. Given what we currently know about the role of serotonin and crayfish social behavior, a specific chemical signal in the urine of the crayfish appears to be powerful enough to alter the inherent social status of the crayfish.

This is similar to the "power tie" mentioned at the beginning of this chapter. This tie not only reflects dominance in the wearer but also forces those individuals around the wearer to become subordinate without consciously realizing it. Imagine attending a potentially hostile business meeting where critical contracts or key business decisions are at stake. Furthermore imagine that you have the human clothing equivalent of the crayfish dominance signal. Simply slip this "super" power coat" on and by the end of the meeting all eyes are on you and ready to follow your lead.

6.4 What's the Password?

Around the age of 10, a few of my neighborhood friends and I formed one of those typical childhood clubs. A club is nothing without a clubhouse, or at least we thought at that age. So we embarked upon that great childhood journey of building a clubhouse. Gathering bits and pieces of old lumber, nails, and screws, we began the task of building our dream clubhouse. If I remember correctly, this was going to be the palace of tree houses with a secret trap door, rope ladder, lighting, and several booby traps for those unwanted visitors. As with many other youths at this age, the mind projects images and ideas that our bodies often cannot see through to the end. Our fabulous tree house ended up as a group of boards slapped together that barely kept out the rain, but to us that tree house was our own private little kingdom.

Another key aspect of a club is having a membership; some belong and some do not. Being young enough that I still had disdain for girls, our club was quite exclusive and stereotypical in that "no girls were allowed." To ensure that only the privileged ones entered the clubhouse, we had a series of top-secret passwords and handshakes. Without performing the correct ritual in the correct order, there was no admittance into our hallowed haven. Looking back however, I am not too sure that any of the neighborhood's opposite sex really wanted to join our club, anyway.

The point, however, is that without the ability to identify those that belong to the guild and those that do not, there is no opportunity for group behavior to arise. Even though I am a fair bit older and hopefully wiser now, I belong to a number of different groups that require a modern type of "password" in order to identify those that are in and those that are out. As a faculty member of the University, I have a special parking pass that allows me to park near my laboratory. My research interests and research papers identify me as a member of the behavioral group within my department.

Outside of my work, I root for a number of different sports teams and during their games I wear the respective team jersey. I used to go to a conference in Florida during the Stanley Cup playoffs, and several years ago my favorite Red Wings were heavily favored to win the Cup again. When I sat at a bar watching the game in my red jersey, I was immediately identified as part of an impromptu gathering of fellow fans. We rooted, cheered, and celebrated the various victories of my beloved Wings and their captain Steve Yzerman. During this time period, the Colorado Avalanche had been their dreaded rivals. Every once in a while, a fellow bar patron would wear their team jersey and some friendly banter was traded between us during a Wings/Avalanche game. These group gatherings and the resulting behaviors were never planned, but arose out of our ability to identify those individuals that belonged in our circle (Wings fans) and those individuals who belonged in a different circle (Avalanche fans). Whether we possessed an ID card, wore a team jersey, or put a lab group on a research paper, these signals sent a clear message of where I belong and do not belong (Fig. 6.2).

Humans, in many ways, are the quintessential social animals. For many of us, we need or even crave to be in a cohort and are unhappy unless there is some association with others of the like mind. In addition to the positive social interactions that come from being in a group, we like being identified as part of a specific group of people. A badge of belonging that many of us wear with a deep sense of pride. Our badge could be a Shriner's hat, military uniform, or a bindi, but all of these are symbols of an exclusive group membership that communicate that fact to everyone that recognizes the symbols.

Fig. 6.2 Ants and insects in jersey

Nature, too, has its group membership and symbols of group membership that can best be seen within the social insects, particularly the social honeybees. Social honeybees have a strong hierarchical relationship and an interesting division of labor. Bees are separated into groups with varying numbers. At the top of the whole hive is the queen. The queen, true to her human counterpart, has numerous workers that busy themselves taking care of her every need. Workers clean her, bring her food, and perform all of the daily needs for the queen. Her only job for the hive is to produce eggs that will grow into future young. The young bees work solely on nest activities, taking care of the larvae, building and repairing the nest, and other housekeeping duties. After a month, they graduate to guard duty. Standing sentry at the door, they make sure that only those bees that belong in the nest are granted entry. Usually after 2 weeks of guard duty, they take the final step and graduate to foraging responsibilities. These bees are the ones we see flying from flower to flower gathering the essential goods for the hive. Our story of passwords and special admission returns to the second stage of duties for social bees: guard duty.

Just as some cities are richer than other cities, some beehives are better off than other beehives. The richer hives have more larvae and more food supplies. Whenever there is an inequity of resources, bees may steal what is needed as opposed to earning a good day's pay by foraging. In harsh times and sparse food supplies, bees will often seek out richer hives and attempt to steal some food or larvae. Thus, the role of the guard bees is to ensure only those bees that belong to the nest are allowed access to the inner sanctum.

Similar to my childhood clubhouse, bees also have a password that signifies those that belong and those that do not. Dr. Michael Breed at the University of Colorado is an expert on social bees and their chemical passwords. In addition to studying the tropical thieving ants introduced in Chap. 1, Dr. Breed has spent a lifetime investigating the chemical language of bees and has noticed some fascinating behavior among bees upon return from their foraging trips.

Foraging bees, as most of us are probably familiar, often visit many flowers in the hope of gathering as many resources as possible for the hive. Laden down with nectar and pollen, they return to the hive only to be stopped at the entrance of their home. Guard bees step forward and perform a thorough inspection of the incoming bee. Tapping them with their antennae, the guard bees chemically inspect the returning forager for their chemical password. If the forager has the right password, they are granted admittance to the hive. However, if they have the wrong password, the guard then summons other guards and a vicious attack is launched on the would-be intruder. (In some cases, the intruder is allowed to pass through without the proper chemical if the intruder offers a valuable bribe in the form of sweet nectar). Dr. Breed noticed that most of the guards would inspect the returning foragers with their antennae, repeatedly tapping and touching various parts of the forager's body. He suspected that the bee was sniffing for a specific signal and proceeded to delve further into this story.

There are a number of different explanations or possible scenarios for the specificity of the chemical password. The passwords could be unique for the individual bees or could be unique for the hive. If the password is unique to the individual, this would be akin to the social security number for Americans. Imagine if you were suddenly placed in the bee's world and are sitting at the entrance of the hive. A forager hovers over to the nest, heavy with pollen and nectar. A guard bee steps forward and asks for their password. In the individual password scenario, the guard bee would have to remember each of the tens of thousands of foraging bee passwords. "Is this Jen, Christy, Sarah, or Patty?" the guard would ask herself. Since bees are only guard bees for 2 weeks, they would have to learn and remember each individual foraging bee within that brief period. Without the help of a computer system for tracking individual records, a specific chemical signature for each individual bee is highly unlikely. If a chemical password for the nest as a whole is used then it could be comparable to the passport system that signifies one's country of origin. This would mean that the guard would only need to learn one password. Therefore, Dr. Breed turned his attention to the usage of a global password that all the bees from a single hive would have.

Like many of nature's products that human industry attempts to copy, bee's wax remains far superior to the wax that we can synthesize. The wax has some interesting properties of strength and pliability, but more importantly for our story is that each hive's wax is unique. Bee's wax is a mixture of the specific genetic make-up of the bees (each hive has a slightly different wax) and the raw materials available for producing wax, which come from the diversity of flowers in the bee's local neighborhood. The key waxy components of a beehive tend to be fairly constant across different hives, but what varies among hives are the minor components that add distinctive odors. A hive located in a field of clover will have a different smell than those hives located in a field of mixed wild flowers. Thus, each hive has an exclusive hive odor that is caused by the local flora and the individual genetics of the colony.

This hive odor is readily transferable to any bee simply by spending time in the hive. To demonstrate this, Dr. Breed performed a series of elegant experiments in which he placed a piece of hive in a small cup with a bee from a different hive. After 5 minutes, he then returned the bee to its original home colony where that "altered" bee was promptly attacked by the guard bees and summarily dismissed as an intruder. If he placed this bee in the colony from which he obtained the piece of the hive, the bee was welcomed in as a loyal citizen. Interestingly, he did similar experiments with the guard bees. He found that a guard bee performs their duties not by memorizing the hive's chemical password, but by simply comparing an odor template of what they smell like to what the intruder smells like. So when a bee returns from working in the fields, the guard greets the worker and takes a good whiff of the incoming bee. The guard then compares this sample to its' own smell and if the chemical passwords match, admittance is granted. Dr. Breed also tested this hypothesis by changing the guard bee's reference smell by placing the guard bees on a bit of bee's wax from another hive for 15 minutes. After this exposure, he then presented the guard with two potential intruders, one of its' own hive and one from the hive that kindly donated the hive wax. The guard bee treated its' own hive member as an intruder and the intruder as an exclusive member of the hive. This password system works well because all of the bees start their lives inside the hive and are constantly exposed to the odors of the home colony. Once a bee leaves that individual will still have the hive's chemical signature upon its return. The chemical password signifies that the bee was already on the inside and deserves admittance. This system seems far more effective than our series of complicated handshakes needed to gain admittance to our exclusive clubhouse. In this manner, the chemical password of the beehive maintains a much better regime of those that belong in and those that need to be kept outside.

As expected, whenever there is an exclusive club, there are those who want in but are not allowed in. If this exclusivity is taken to the extreme and there are differences in resources for those animals in the club as opposed to those outside the club, the exclusivity can lead to potential conflict. In nature, this conflict can escalate to hive, troop, or intergroup warfare. Whether the conflict is over admission to private territories, conflict over mates, food resources, or shelters, or over the life and death struggle of predator–prey interactions, chemical signals are playing a role in some of the most intense struggles found in nature.

6.5 With a Little Help from My Friends

Despite being somewhat of an introvert, I have spent a good portion of my life as parts of different teams or groups. As mentioned above, all of these teams have different badges or symbols that acknowledge some type of group association. Many years ago, I played numerous sports in high school. Each team had a style of clothing, coloration, and mascot that would serve as some sort of rallying focal point. Academic and band groups also had points of focus that would bring people together for a common cause. Now as an adult, I find myself in many similar group or team situations where individuals have come together to achieve certain outcomes.

Within the ranks of faculty at a modern university, faculty committees are formed, given a name, and charged with accomplishing certain tasks around campus. I am a member of a martial arts club, and our symbols and rituals are designed to foster an atmosphere of cooperation and support. Communities can be built around symbols and rituals.

Certainly, within the sciences, individuals can accomplish their goals of performing good science, but science is inherently a social endeavor. Papers and presentations are reviewed in a social manner, graduate students pass through different public or social hoops in order to signify their level of readiness to be a scientist, and in fieldwork, cooperation and help are critical. In many ways, scientific survival is dependent upon cooperation and teamwork. Probably the group that I am most proud to be a part of is my group of graduate and undergraduate students that comprise my lab. This group, called "The Laboratory for Sensory Ecology," is a large, boisterous group full of strong personalities. Every year we produce t-shirts with the lab logo or some other drawing/saying that is key to that year. In addition, when students have significant accomplishments, we gather and celebrate them as a group. For paper acceptances, grants, and graduation, the students get their own bottle of champagne. As the group gathers, the celebrator globs some paint on the top of the champagne cork (the color of the paint symbolizes the event), pops the cork, and marks the ceiling. Finally, the student climbs a table and signs their name on the ceiling next to the paint mark. We all cheer and congratulate them on their accomplishment. These badges (t-shirts) and ceremonies (popping champagne) of belonging serve to draw the social group together and foster an environment of collaboration.

While completing his work "In Memoriam" in 1849, Alfred Lord Tennyson created the phrase "…Nature, red in tooth and claw" to symbolize the often violent nature of competition and natural selection as opposed to the collaboration described above. Some of the stories about hierarchies described above might also give rise to the notion that nature is often full of conflict and warfare. Cooperation also exists within nature and that cooperation occurs across different levels of groups. In birds, there are cases where the previous generation of siblings will stay to help raise the next generation of brothers and sisters. The Australian mudnestors are obligatory cooperative breeders meaning that without helpers the parents cannot fledge their offspring. Cooperative breeding also occurs within mammals such as Meerkats and some primate species.

Beyond breeding, cooperation is seen among many different groups for predator protection. Among Canada geese, individuals will take turns being vigilant for predators while other geese are foraging. Two to three geese in a large flock will voluntarily stop eating in order to stretch their necks looking for potential predators. After some period of time, these geese will begin foraging and a couple others will start their turn watching for predators. This rotating set of eyes (rotating turns for guard duty) allows the flock to increase their overall efficiency for foraging, while maintaining a safe level of detection for predators. There isn't any active communication about coordinating this behavior among the flock, but the group behavior arises from the periodic turns that animals take being vigilant. The same type of vigilant behavior is commonly seen for prairie dogs, zebras, and other savanna herbivores.

In some instances, the coordination of group behavior is performed through communication, chemical, or otherwise (as seen with Dictyostelium aggregation outlined in Chap. 1). In primate communities, vocalizations are used to signal the presence of predators, and different sounds are used to signal danger from above (raptors), in the trees (snakes), or from the ground (jaguars). Communication to coordinate group behavior is seen throughout nature and not just in animals or bacteria. Even plants will exhibit a rudimentary form of communication between members of a "group." The term group here is used quite loosely because there isn't any active behavior on the part of the plant to group themselves. For the most part, grouping is done through the varied processes of reproduction and seed dispersal.

Plants are not the first group that are commonly thought of for examples of behavior and communication, yet this group of organisms provide excellent examples of chemical communication. As mentioned in Chap. 1, plants can produce a number of chemicals that are sequestered in the leaves and other tender parts of the plant that are of particular interest to herbivores. In a previous example, oak trees would release a chemical signal through their roots that served to increase the anti-herbivory chemicals in the leaves of surrounding oak trees. Given that roots are designed to release and take in chemicals from the soil, this example may not be too surprising. What may be more surprising is that plants also send signals through the air to other plants to communicate predatory events. Acacia trees in Africa are a source of nutrition for the numerous herbivores on the African savanna. These trees, once attacked by an herbivore, can produce chemicals and sequester them in their leaves. The chemicals, a class of compounds called tannins, are aimed at reducing leaf herbivory and are so toxic at high enough concentration that kudus have been known to die from an overdose of toxic leaves. Another prominent African herbivore, the giraffe, has been observed consuming the leaves only on upwind trees and avoiding the luscious leaves of downwind trees. In addition to producing more tannins in their own leaves, the attacked tree also sends airborne chemical signals (pheromones) to the surrounding group trees. As the downwind trees detect the pheromone, they start producing more toxins and these leaves become, at a high enough concentration, lethal to the kudu.

Some acacia trees don't stop with these simple chemical defenses though. In a true sense of calling on friends for help in times of need, the acacia trees are master alchemists. Now, the concept of chemical communication in plants was not an easy one for the scientific community to accept. Early pioneers in this field, such as Dr. David Rhoades at the University of Washington, continued work on a wide variety of plants and their chemical reactions to insect attacks. The concept of chemical communication between plants of the same species has become fairly common and even some evidence on communication across different species of plants. The bullhorn acacia, found in Central and South America, has a unique relationship with ants that helps protect the plant against the relentless attacks of grazing herbivores.

Acacia trees in general have thorns that grow on the branches that are part of their deterrents to grazers such as antelopes, but on the bullhorn acacia, the thorn has expanded and been enlarged through natural selection. The thorn still functions as an anti-herbivory defense, but is also a home to an ant (*Pseudomyrmex ferruginea*) that has a symbiotic relationship to this particular acacia. In symbiotic relationships, both organisms benefit from their close relationship. In some instances, the organisms involved in the symbiotic relationship cannot survive without the other organism. With acacia trees and this species of ants, the tree provides the ants with precious food produced at the tips of the thorns. The ants, in turn, provide the trees with protection against those organisms that would like to consume the tree's leaves. Unlike the acacia trees in Africa, this particular species of acacia doesn't produce the tannins in response to grazing on its leaves. Work performed by Dr. Anurag Agrawal (Cornell University) found these trees have the ability to "call" for help from the ants that live within the thorns. Upon a disturbance (break of leaves or branches or grazing), the acacia tree emits a special pheromone that attracts the ants to any spot of damage. In one set of trials, Dr. Agrawal punched a hole in a leaf with a common paper hole punch. Within 4 minutes the number of ants patrolling the damage site had increased by 400 %. The ants, riled up and gathered in one spot by the chemical distress call, will attack any herbivore in the area and protect the tree from any further damage. Having the right friends in the right places at the right time definitely pays off; of course, having the right pheromone is necessary to place that call for help.

6.6 The Life of the Party

In Chap. 2, I described the neural connections for our sense of smell and compared them to the neural pathways for our vision and hearing. To summarize again, the information from the eyes and ears go to specific areas in our cortex and as such, we think about these stimuli before responding to them. The thought process might be exceptionally quick or a slow deliberate process. In contrast, the information from our nose is sent to the limbic system which evokes a powerful emotional response. The delight of the smell of food, the sensual nature of a sinful perfume, or the refreshing aroma of air after a morning thunderstorm all provide strong sensations. Within a social context of the human existence, I would define our olfactory senses as rather judgmental. The immediate emotional response, rather than some cognitively developed response, evoked by chemical signals colors our perception of people in social settings.

Another thought experiment might provide some illumination on this concept of judgmental olfactory responses. Imagine you are off to a meeting; say, a fairly important meeting. Definitely not a business casual meeting, as there is an unwritten expectation that everyone should look "presentable." The meeting occurs in some high rise building that requires an elevator ride. Maybe the elevator is slow or the building is tall, but there will be a substantial period of time riding in the relatively small and closed space inside of the lift. As you enter, a rather well-dressed middle aged man follows you onto the elevator. Perhaps, he is going to the same meeting that you are attending. As the doors close, you perceive a subtle hint of body odor. There are only two of you in the elevator and you showered thoroughly in the morning, so clearly this aroma is arising from your companion. As the ride gets longer and longer, the odor increases in intensity. Without much thought, what are your conclusions about this gentleman? What would change about those judgments had the gentleman smelled pleasantly?

The scientific work in the use of smells in judging others is still relatively sparse, but Drs. Chuck Wysocki and George Preti at the Monell Chemical Senses Center in Philadelphia have spent years collecting, analyzing, and testing different bodily odors that are produced under different conditions. Humans have a tremendous ability to produce odors and these odors appear to be different given the condition or mood of the odor donor. Human sweat will have a variety of smells depending on whether the donor is scared, excited, or sexually aroused. Although most of the work has focused on the axillary glands located in the armpit, there are other sources of human odors such as the mouth, feet, and genitals. Drs. Wysocki and Preti have categorized some of these odors as either primer pheromones (those that impact the endocrine or nervous system), releaser pheromones (those that evoke a behavior), modulator pheromones (those that impact the moods of other people), or signal pheromones (those signals that supply specific information such as reproductive status or sex of the sender). Within a social context, the modulator pheromones may be the most interesting because these chemical signals may influence the mood of the social environment without the other people actively noticing or being aware of those signals.

The work by the famous psychologist Martha McClintock has shown that the odors emanating from lactating women have the ability to increase the sexual motivation in other women. Interestingly, this effect was different for women with a regular sexual partner as opposed to those women that were single (the key to being single is not related to sexual activity but connected to companionship). The increased sexual motivation was higher in those women with partners. Dr. Preti and Wysocki have shown that women's sexual moods were increased if they were presented with odors from the armpits of male participants. These are just two examples of how chemical signals can alter or modulate the moods of people without any cognitive recognition of those odors. I would like to refer back to the definition of pheromone covered in Chap. 2 and to note that pheromones and, in particular, these modulator pheromones alter or enhance moods of people on the receiving end of the odor. The search for a human pheromone that evokes uncontrollable behavior as depicted in popular culture is probably a fruitless search.

These studies show that social judgment based on odors is possible within humans. In the scenario above, the body odors produced by the elevator companion may tell us a lot about the emotional state of the gentleman. Work by Drs. Denise Chen and Jeannette Haviland-Jones has demonstrated that humans are capable of detecting different moods of people based on their body odor. In this study, participants were asked to watch either short clips of a happy or comedic movie and other participants were asked to watch a short clip on bugs, spiders, and snakes. Collecting human body odors doesn't sound like a fun job, but there is a standard technique that is relatively straight forward. Those people watching the movie wear cotton pads underneath their arms and then the cotton pad is used to produce odors. At some later point, both the donor and naïve people are asked to come in, smell vials of body odor, and asked to identify whether the odor came from a happy person or fearful person. The participants could accurately assess the mood of the donor person. A more recent study, led by Dr. Gün Semin, showed that people who are exposed to context specific body odors tend to have the same emotional response as the donor person. For example, Dr. Semin showed male participants one of two different movie clips to evoke specific emotional responses. They were shown either horror clips (*The Shining* with a wonderful performance by Jack Nicholson) or disgusting clips (scenes from MTV's *Jackass* show). As in the studies above, cotton pads were placed underneath the armpits of the donor males. Females were invited in to take a whiff of the odor, and the researchers tracked eye movements and facial muscle contractions in order to capture their initial emotional responses to the chemical stimulation. (Males were selected as donors because they produce stronger chemical signals, and females were selected as tester because, on average, females are more sensitive to chemical signals.) The researcher found that the female participants reacted in an identical fashion as the emotional context of the odor. In other words, if the participant smelled a "fear" odor, they reacted with a fearful expression and those participants that smelled a "disgusting" odor reacted with disgust.

The authors have labeled this phenomenon an "emotional contagion" where the emotional state of one individual may spread to others via chemical signals. To extend this analogy farther, scary movies may be even scarier in crowded theaters where some individuals may be (unintentionally) sending fearful signals across the theater through body odors. If the gentleman that entered the elevator with you is in a state of fear, perhaps because of the impending big meeting, you may also begin to feel a heightened sense of fear after "receiving" his chemical signal. Similar to the bullhorn acacia trees, maybe this is our response to potential danger and that chemical signals (as modulators of our moods and behaviors) could be a powerful and hidden part of human society. In the next chapter, I'll explore how these social signals could evolve into something a little more sinister as animals have developed the ability to lie, cheat, and deceive through their odors.

Chapter 7
Stealth and Deception

Lying and Hiding with Chemicals

During those mornings that I have appointments off of campus, I often listen to National Public Radio or BBC America on the way to these meetings. The style of reporting aligns seems to resonate with how I like to hear the news. Typically, politics and world events dominate the news stories and the interviews that are played during the morning hours. Straightforward and probing, the reporters asking the questions seem to understand that the initial responses of those that they are interviewing are not always truthful. Maybe I have a cynical view on politics, but I am doubtful that any politician, independent of country or political party, is adept at telling the truth. Within American politics, lying appears to be an art raised to its highest form. In an excellent book on his life's work called "Telling Lies," Dr. Paul Ekman outlines several types of lies that range from small to large and from whether the liar actually believes that they are telling the truth versus the recognition that everything is a fabrication. Another point that becomes clear in Ekman's writing is that we all engage in lies at some point in our lives. Even the smallest fibs or those told with good intentions are considered lies at some level. For example, if one feels a cold or flu coming on, but doesn't want to alarm their spouse, they could respond with a hearty "I am fine" when queried about their health. We would consider this harmless or maybe even beneficial because the liar wouldn't want to add to the stress of the listener.

At the other end of the spectrum from this harmless little falsehood appears to be the daily tails spun by politicians when they are campaigning or speaking to the media. One of my pleasures at this moment in time is watching a superbly acted Netflix series "House of Cards." The two lead characters, played by Kevin Spacey and Robin Wright, are Washington insiders and politicians who play the game of deception fairly well. The series title refers to the layers of deception that are built up over numerous interactions and upon which these two characters are precariously perched. The house can collapse at any moment in time if a single layer of lies is exposed. If that house fell, the world built by Spacey's and Wright's characters would also fall very unceremoniously. The series is excellent and I do, at times, get a sense that the portrayal of large-scale Washington politics and the deception

P.A. Moore, *The Hidden Power of Smell*, DOI 10.1007/978-3-319-15651-4_7

contained within those politics is far too close to reality. As political power, money, and fame are at stake with every word or action that the staffers and politicians take, the drive for the use of lies is far too powerful to avoid.

Within the politics of nature, power and fame are not at stake, survival and reproduction are. As in our world of politics, nature has its own fair share of deceivers and liars. All organisms need resources. Whether its plants using nutrients for growth or animals seeking shelter for protection from the elements or predators, resource acquisition is an important part of the daily lives of organisms. Most resources in the natural world are limited, and thus, there is intense competition among animals and plants for the critical nutrients and resources needed for survival and reproduction. Although there are examples of animals cooperating in order to achieve the desired outcome of resource acquisition, most of nature is locked in a deadly competition. Within this struggle for life, organisms will cheat each other out of resources, steal, and use any means possible to get just a little bit more of the survivorship pie than their competitors. This is evolution.

Charles Darwin, when writing his manuscript that would become "On the Origin of Species by Means of Natural Selection," was heavily influenced by the Scottish economist and moral philosopher, Adam Smith. In Smith's influential book "The Wealth of Nations," he conceived of the invisible hand that guides the development of a well-ordered economy as individuals compete for financial success. In this case, the success is defined as wealth. For Smith, the invisible hand was not a set of strict obvious rules, but an unwritten set of business principles that arose naturally as companies competed in a free and open marketplace. Those companies that produced their products of high quality would survive the economic battle. The concept of the invisible hand meant that there wasn't a need for an oversight body that regulated companies. Competition would be the regulating force such that if you delivered the best product for a fair price, that company would survive. Darwin, while he was attempting to make sense of his many observations from his 5 year voyage on the Beagle, had the brilliant insight to apply Smith's concept of competition for wealth to nature. For Darwin, animals were also competing, but instead of wealth that produces a well-ordered economy, they were competing for reproductive wealth: that is leaving as many offsprings as possible. As animals competed, a well-ordered ecology would naturally arise without the need for an external regulating force. Just as in the "House of Cards" political drama or real world politics, there are no rules in competition. Organisms are free to cheat, hide, and deceive in order to outcompete others, and chemicals and chemical signals can be a tool that allows deception to occur.

7.1 It's a Trap

A spring stroll through any flower garden demonstrates one of the best examples of nature's "negotiated" treaty between insects and plants. This peace treaty is called coevolution. In actuality, the insects are not the only exclusive members involved in this treaty as bats, ground mammals, birds, and even lizards take part. All of these animals function as pollinators, and during their foraging runs between plants they carry with them the pollen of the flowers that they have previously visited. With each new visit to a plant, these animals deposit pollen, gathered from the previous plant, to ensure that these plants will reproduce. Now, this group of pollinators don't perform this service *pro bono* and demand some sort of payment for their work. The flowers produce this payment in the form of high octane nectar. The nectar is what draws in most of the pollinators (although some insects will use the pollen itself), and in the process of getting at the nectar, these animals inadvertently transfer pollen from flower to flower.

As these two groups (flowers and pollinators) evolved through time, they became dependent upon each other. Bees have no other source of food and are entirely dependent upon the flowers for food. The flowers would not be able to pollinate each other without the visitations of the bees. Some species have evolved such a tight relationship that there are specific structures within the flowers that allow specific pollinators to do the job. Hummingbird flowers are such a group with adaptations that, as the name suggests, cater only to hummingbirds. Aloe Vera and Desert Honeysuckle are two such flowers and typically have long tubular shaped blossoms. Hummingbirds have long thin bills that allow them to drink the nectar at the bottom of the blossom. In addition, hummingbirds have visual systems that are tuned to specific colors and many of the hummingbird flowers have the red, orange, yellow, or blue flowers that activate the visual system of the birds. Without this coevolutionary adaptation (or peace treaty) between bees and flowers, we would not have beautiful gardens that surround our houses or the sweet honey for our toast and tea. Coevolution, in this case, is the result of natural selection working simultaneously on two different species where the change in one (e.g., position of the pollen in the flower) results in a change in the other (e.g., increased hair length on the legs of bees). The previously mentioned hummingbirds and flowers are an excellent example of coevolution. Flowers that are pollinated by hummingbirds (honeysuckles or torch lilies) typically have a trumpet shape to their flowers, and the nectar is located at the bottom of this trumpet. As the trumpets evolved to be longer and longer, only the hummingbirds with long beaks could feed. Other insect pollinators couldn't receive their nectar reward for pollinating, so they stopped visiting these flowers (over evolutionary time).

Similar to human treaties, there will always be some groups willing to use the rules of the treaty in order to get an advantage. Usually, the advantage involves something like natural resources such as access to oil, gold, diamonds, or exceptionally good fishing or hunting grounds. Nature too has its share of members who have broken the treaty. In the case of the coevolution of flower and pollinators, the flowers advertise their delicious meal of nectar through chemical signals. As mentioned in Chap. 3, the flowers are advertising their home cooked meal trying to attract in the visitors that will help with the pollination. The nectar's aroma signals the presence of a reward for a passing insect and all that the insect has to do is come inside the flower's restaurant.

In nutrient poor soils, a large group of plants have evolved unique ways to obtain the necessary nutrients in order to survive, and they have done so by deceptively

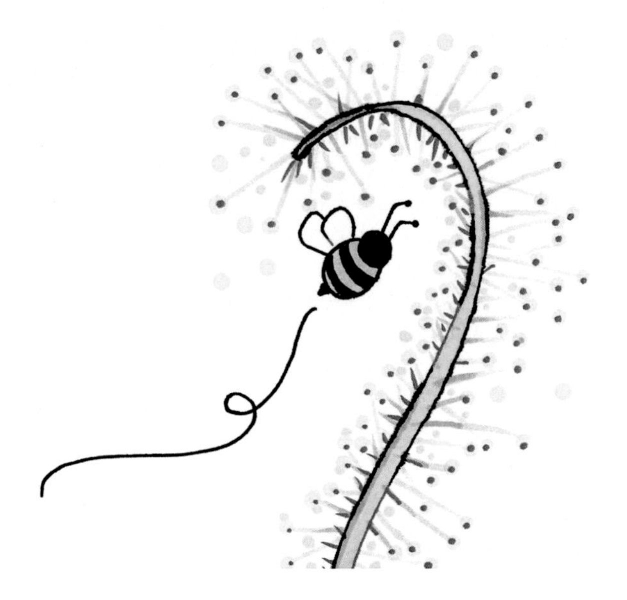

Fig. 7.1 Sundew meal

playing the role of a flower looking for a pollinator. The sundew plants, in the genus Drosera, are a large group of species that have evolved stalked glands that cover their leaves. The name sundew originated because the tips of the stalked glands glisten as if they are covered with drops of the morning dew. Yet, the drops are not condensed water, but are mucilage, which is a sticky gooey substance that the plant secretes. The mucilage smells sweet and tricks the passing insects into thinking that there is a nice free meal of nectar just waiting to be picked up. An ant or small fly that is unlucky enough to be tricked will approach the glands and lightly land in order to gather their reward (Fig. 7.1).

As soon as the insect touches the stalked gland, a series of surprisingly quick reactions occur. The stalks surrounding the one that was touched will quickly bend as the entire leaf contracts upon itself. The insect is then quickly contacted by sticky stalk after sticky stalk. This would be similar to placing a penny in the palm of your hand and closing finger after finger upon the penny. The difference is that we only have five non-sticky fingers, and the sundew can have dozens of sticky fingers that quickly entrap the helpless insect. As each stalked tentacle contacts the insect, more and more of the mucilage covers the prey until the entire animal is enveloped in the sticky substance. The insect either dies from exhaustion or is suffocated to death after being covered in the gooey substance.

At this point in time, the sundew excretes enzymes to slowly digest the insect to get needed nutrients. In the nutrient poor environments where the sundew thrives, evolving mechanisms to extract a nutrient source not used by other plants gives the sundew a selective advantage. The sundew produces chemical signals that have hi-jacked the coevolutionary pathway used by insects and flowers. Usually, the sweet smelling signal means a reward for the pollinator for a little bit of extra work.

The sundew signal chemically mimics this same signal, and thus, the insects are drawn into the deadly trap. Here the lethal deception performed by the sundew is the last thing that the insect will ever smell.

7.2 Trick or Treat

In northern Ohio, the cool and crisp evenings coupled with colorful leaves falling from the trees signal that it is time for the annual October 31st ritual of trick-or-treating. Our house lies just outside the main drag for trick or treaters, but we still get our fair share of visits to the house. In our local area, we have set times when the official Halloween activities begin, and promptly at 6:30 pm the knocks on our door start. Throughout the night, a good number of ghosts, vampires, pirates, and a string of the latest animated movie characters knock on the door and yell “Trick or treat” as we pull open the entrance. Sometimes, the parents come forward with the youngsters and appear dressed in some family themed costumes.

What appears to be particularly favorable among the many visitors is some Disney character. Since the recent explosive hit movie was *Frozen*, many of the little girls are dressed as Elsa or Anna. White braided hair, long flowing blue dresses, and sparkle makeup are the key elements of the costume for those wanting to be Elsa. The girls are attempting to mimic the look of the Disney queen and younger sister. The trick or treaters do a good enough job that we can tell which Disney character they are attempting to emulate. Carefully chosen elements from the original character’s look and dress are used to create the illusion of a mimic. The three elements (dress, hair color, and makeup) listed above are most commonly associated with the original Elsa. Whereas choosing to mimic Elsa’s shoes, eye brows, or nose probably wouldn’t lead to a successful illusion.

Within the natural world, mimics are organisms that send a litany of sensory cues designed to lie to any potential predators. Through natural selection, these cues have been honed in such a way that to distinguish between the real organism and the mimic is exceedingly difficult. The most common mimics and those most studied by scientists are visual mimics. The classic example of nature’s mimics includes the monarch and viceroy butterflies. Monarch caterpillars feed on milkweed plants and sequester alkaloids within their body tissues. The alkaloids remain in the adult butterfly’s body. When the adult monarchs are consumed by birds, the birds get sick and regurgitate the butterfly. Although that poor butterfly dies in the process, the bird learns to avoid the characteristic orange and black of the monarch. By mimicking the color pattern of the monarchs, the viceroy, even though it doesn’t have the toxic chemical, also avoids predation.

With the trick or treating Disney queens and princesses, there is no selective pressure for the mimic (the little girl) to be as close as possible to the model (Elsa). Within the viceroy and monarch mimicry, there is selective pressure for the viceroy to be as close to the model as possible. Monarch butterflies are unpalatable if the caterpillar feeds on milkweed during that stage of its lifetime. The caterpillar that sequesters alkaloids from the plant which makes both the caterpillar and adult butterfly unpalatable. Avian predators of the adult butterflies learn of the unpalatability after just a single attempt at consuming the monarch butterfly and learn to avoid any and all butterflies that appear similar in coloration. Mimicry is just one of nature's deceptive tricks.

Apart from the host of colorful costumes that parade across our porch in search of candy, there are also costumes designed to conceal the wearer. Several little batmen dressed in the midnight black version of the costume appear as well as a couple of little ninjas. The ninjas are dressed completely in black material except for the thin strip of exposed skin around their eyes. Both costumes, in their original form, are designed to blend into the background. Batman, active at night, wears a dark costume in order to hide from potential criminals. The ninjas chose black for the same reasons as they made most of their attacks during the time after the sun was set. A third set of costumes can also fit within this type of deception and that is of the little soldiers or hunters dressed in camouflage. The theory behind the design of these visual images is to allow the wearer to blend into the background whether that background is the dark of night or the forests for the hunters. The visual signals are selected to conceal rather than mimic an existing organism as in the viceroy. This type of natural deception is crypsis, formally defined within the scientific literature as the ability of an organism to avoid detection by other organisms. In the deception outlined above with the butterflies, the deceiver (the mimic) needs to be obvious to other organisms for the deception to work. The mimic needs to be seen and mentally compared to the model or else the deception is useless. In crypsis, the organism is attempting to go completely unnoticed.

In nature, there are many examples where both predators and prey use crypsis for an advantage. Many ambush predators use crypsis to remain completely hidden until unwary prey approach within striking zone. A beautiful example of this is the bark mantis. This brown, white, and green praying mantis remains motionless on a tree until a poor insect falls victim to this deception. However, maybe nature's champion at active crypsis is the octopus. With numerous chromatophores and an endlessly pliable body, the octopus can assure a wide range of shapes and colors in order to disappear into its background to avoid predation and act as predator.

Jumping from the younger activities at Halloween to more adult endeavors during the carnival season, trick-or-treating can be replaced by seventeenth and eighteenth century masquerade balls. Participants would arrive in elaborate and fancy dress and a simple mask to hide their identity. Party goers would revel in anonymity until midnight when the masks were removed and the identities revealed. In this type of deception, the wearer of the mask is not attempting to mimic or to hide as in crypsis, but is simply "masking" their identity to remain unidentifiable. Although not as common of a deception as mimicry or camouflage, masking does occur within nature's world. Typically, masking occurs when the intended signal, the bird song, or frog croaking is lost among the background noise of the forest. The term masking refers to one stimulus (background noise) that is of higher intensity than the other signal (bird call). In human terms, this would be equivalent of attempting a conversation at a masquerade ball while the string quartet is playing a rather loud waltz. In this example, there are two signals being masked (the facial features and the voice). Sometimes, these masquerade balls would end in tragedy as assassination attempts were not uncommon given all of the deception that was occurring. One could imagine a ball where all three types of deception were present if two assassins were hiding in the shadows (crypsis) attempting to kill the king who, knowing that an attempt was likely, had a soldier dress in his regal attire (mimicry). To coordinate their attack, the assassins used hand gestures that were unseen (masked) amongst the movements of the waltzing patrons. In similar fashion, many animal assassins hide their appearance in order to perform their deadly duty.

7.3 The Pirate's Ghost

Despite some of the more elaborate costumes that come by the house during the Halloween ritual, the most common and probably easiest to construct is the classic ghost costume. Simply take a sheet, cut out some eye holes, and drape the sheet over yourself or your child. In one of my favorite comic series, Peanuts, Charles Schulz made a common practice of dressing up the young children in a simple ghost costume during many of the funny Halloween panels. In typical Charlie Brown fashion, the young boy has trouble with the scissors and instead of two eye holes, has a rather large number of holes. The costume is still very recognizable despite the numerous holes.

The concept of a ghost costume as a simple sheet works in large part because of the collective meme that society has about the shape of ghostly objects. Ghosts, specters, or apparitions are described as amorphous blobs hovering slightly off the ground usually emitting an eerie moan or wail. The Halloween version of the ghost with using a sheet ignores a commonality that ties together almost all definitions of ghosts and that is transparency. The visually deceptive nature of the ghost is the invisibility or just barely visual nature of the specter. This visual aspect is so prevalent to the word ghost that the English language has co-opted the term for uses outside of the supernatural. As whiteboards fail, they "ghost" or leave faint traces of the previous words written with dry erase markers. For a literary turn of phrase, one could claim that there was a ghost of a smile after a particularly subtle joke. These uses of the term indicate something (a smile or old dry erase mark) that may or may not be detectable upon initial visual inspection.

All of the above examples are visual references as the concept of ghosts are most often associated with our eyes rather than our ears or noses. Most certainly, armies have taken advantage of how our visual systems function to construct effective camouflage such that soldiers and equipment disappear into the background or become unnoticeable. There are numerous examples of animals using these same visual tricks to become "ghosts" in nature; some were mentioned above. The very nature of chemical signals and their production make the concept of a chemical ghost as difficult to grasp as a ghost itself. A chemical ghost would be able to hide in plain sight becoming a mere faint of its former aroma. The very fundamental being of life is chemistry. All of life transforms chemicals from one form to another whether plants take in CO_2 and produce O_2 and sugars during photosynthesis or humans take in food to produce body odors and CO_2. So, to become a chemical ghost appears to run counter to the principles of nature, but recent work has shown that maybe there are really such things as chemical ghosts.

In most instances of chemical camouflage, mimicry, or crypsis, the chemical deception is highly specific, meaning that the intended targets of the deception are rather narrow. In the orchid example below, specific species of bees are targeted as the orchid produces the mimicked sex pheromone. (In contrast, an army's or an octopi's use of visual camouflage is general deception in that any visual organism can be deceived by the cryptic coloration pattern.) The specificity of chemical deception is due to the relative specificity of most odor receptors as opposed to the more broadly sensitive visual pigments in animals' eyes. For generalist predators, seeking to become chemical ghosts to all of their potential prey, producing several chemicals, each chemical targeting a specific prey item, would be a costly endeavor.

The pirate perch (*Aphredoderus sayanus*) is a small freshwater fish that inhabits rivers and lakes along the Eastern and Midwestern part of the United States. This fish is called the pirate perch because of its propensity to ambush and eat other fish when placed in an aquarium with them. The adult fish can reach a length of around 5 inches and is a generalist predator that can consume small insects (dragonfly and stonefly larvae) as well as shrimp and worms. Being a generalist predator has some advantages in that there are always different prey choices around. If one prey item were to disappear from the environment due to disease or some other catastrophic event, generalist predators have many other options for food. The disadvantage of being a generalist predator is that the predator has to hide from so many different eyes, ears, and noses.

Within the aquatic or marine environments, animals and plants can be considered leaky bags of chemicals. Animals and plants need a certain set of ions, nutrients, and fluids to continue to live. This internal living "soup" needs to stay fairly constant or else our bodies do not function correctly. Those organisms living in an aquatic environment do not have to be concerned with losing water to their external environment because whatever that is lost can easily be replaced by water from their habitat. Thus, these organisms have "leaky skins" as water moves relatively easily across their external barrier. In contrast, in our native terrestrial habitats, our (and animal's) skin is an effective barrier to protect against moisture lost. Since we live in a dry habitat (i.e., air), we have evolved a water tight covering that keeps our internal environment fairly stable. The leaky bags of chemicals (or plants and animals) release numerous potential signals for predators and prey to detect the presence of other organisms. A generalist predator, such as the pirate fish, can be considered a leaky bag of chemicals and a wary prey could easily detect its presence. Whereas that same wary prey would be at a significant disadvantage if a predator could be transparent to prey like a ghost.

The pirate perch appears to have evolved an interesting, but still unknown mechanism to become a predatory ghost. To investigate this amazing and unique ability, Drs. William Resetarits (Texas Tech University) and Christopher Binckley (Arcadia University) ran a series of field experiments using a number of different fish including the pirate fish. They constructed a number of artificial ponds and placed within these ponds different types of predatory fish. The scientists placed a different fish in separate ponds such that a single pond was home to a single species of predatory fish. They used sunfish, catfish, and minnows among other predators. Then, in a fairly elegant design, they covered the ponds with very fine mesh netting and sunk the netting below the water's surface. The netting is necessary because this treatment protects any insects that happen to colonize the ponds from predation. The insect colonizers couldn't see the predatory fish in the ponds, but could certainly smell any chemicals being released by the fish below the net. After waiting several weeks, the authors of the study came back to the ponds and simply counted the number of insects that colonized the ponds (mostly beetles but they did count 14 different types of prey). Interestingly, they found that the pond with the pirate perch had significantly more prey in it than the other ponds. The prey were choosing to inhabit the pond with the pirate perch over the ponds with the equally deadly predators. Somehow, the pirate perch had become chemically invisible to the potential prey items.

This is a newly discovered phenomenon within the chemical ecology community as this is the first instance of a predator that has evolved a mechanism that appears to make it chemically transparent to its prey. Hopefully, this experiment is just the start of a new line of research as the underlying mechanism on how the pirate has become a ghost is not yet understood. Perhaps the perch has managed to smell unlike a fish or has developed a way to mask the normal fishy smell that prey detect. Whatever the mechanism, the perch has evolved the ability to perform a chemical disappearing act, the ability to chemically camouflage is, up to this point, unknown within the animal kingdom, and is an exciting line of work. One thing that is certain about this story is that evolution is a far better designer of a ghost costume than Charlie Brown ever was.

7.4 To Bee or Not Too Bee

The hymenoptera is an order of insects that includes wasps, bees, and ants. As detailed in Chaps. 1 and 4, as a group the hymenoptera rely quite heavily upon chemical signals to perform ecological functions. For ants, the recognition of colony members by their body odors is critical in determining who is allowed into the nest and who is forcefully marched off the premises. In wasps, the queen is figuratively the queen bee of the nest and is the only one allowed to mate and produce offspring. She accomplishes this edict by emitting a chemical that suppresses the ability to reproduce in her competing females. Our spring and summer gardens would be devoid of flowers if not for the constant stream of flying visitors gathering a quick meal of nectar in payment for spreading the pollen to other waiting flowers. All of these examples show the highly evolved nature of the odorous world for this group of insects. Without too much of a stretch of the truth, the absence of these chemical signals would signal a very quick extinction of this rather large group of insects.

This last relationship, between pollinators and flowers, is the epitome for coevolution. Within the evolved relationship between flowers and hymenoptera (as well as other pollinators), there exists a quid pro quo relationship. The flowers "pay" the insects with nectar while the insects perform the service of transporting pollen to other flowers of the same species. If one of the two members of this alliance were to fail on their end of the evolutionary bargain, natural selection would act to either produce flowers that continue to provide nectar or produce bees, for example, that found other sources of food. An evolutionary advantage exists for flowers that reward bees and that advantage more visits by the pollinators. Still, within a field of honest flowers (or flowers that reward pollinators), cheaters will eventually evolve. This is nature's equivalent of a Tom Sawyer. The Mark Twain fictional character was infamous for getting other kids to do his chores or labors and even convinced them to provide him payment for the privilege of doing his work. One such group of Tom Sawyers of the plant world are the impressive flowers of the orchid family.

There are approximately 20,000–25,000 species of orchids in existence and in my opinion, they are some of the most beautiful groups of flowers given their unique shapes. The flowers have an immense variation in color and structure and as such are favorites for horticulturalists. One group of orchids have evolved a particularly interesting method for pollination that relies heavily on cheating the pollinators out of their reward. This genus of orchids (Ophrys) are called bee orchids and have flowers that at first glance look remarkably similar to bees. The flowers have evolved a petal that is enlarged and covered with the brown and yellow coloration similar to the abdomen of a bee. The modified petal is termed a labellum, and there are even eye spots located above the labellum to complete the bee costume. The bee-like appearance of the flowers is just the beginning of the ingenious and deep deception performed by this flower. To return to the Halloween example above, the little girl in the Elsa costume is a good mimic, but if the girl could talk and sing like the movie character Elsa, then the deception would be much more complete (Fig. 7.2).

Fig. 7.2 Bee orchid and bee

The orchid's version of Elsa's voice is the sweet perfume that has evolved to be the coup de grâce of the mimicked bee costume. In addition to the floral scent of the orchid, the flower has evolved the ability to reproduce the female mating pheromone of the bee. (A mimic pheromone is called an allomone.) In the case of the orchids, the allomone is such a close mimic to the real sexual pheromone that the male bee actually attempts to copulate with that part of the flower that appears to look like a female bee. The male bee, who has been deceived to believe that the flower is a female bee, picks up pollen and transfers that pollen to the next orchid that draws his sexual interest. At this point, the orchid has performed what is termed a "sexual deception." By tempting the bee with the false promise of copulation, the orchid has actually gotten the bee to do its own sexual work by delivering pollen to its mate. To really complete the Tom Sawyerish nature of the deception, the flower lacks any nectar reward for the bee. The orchid receives its reward (pollination) by tricking the bee into thinking that there is some fun available from the flower. The pseudocopulation by the bee serves to pollinate the orchid to ensure the next generation of orchids. The sexual deception has short circuited the coevolutionary relationship between the pollination work (the trick) and the nectar reward (the treat).

During the Halloweens of my youth, many kids would come to my house and ring the doorbell. If I wasn't out in costume myself, I would rush to the door in hopes of beating my father to greet the kids. If I lost out of this race, my father would respond with a hearty "trick" when they yelled "trick or treat." After a moment's confusion on the part of the ghosts and goblins, my father would require that the kids perform a joke or song (a trick) before giving them candy (a treat). Although embarrassing for me to watch this out of norm Halloween behavior, at least my father was getting something in return to handing out the free treats. As a grown up, I don't require this mandatory trick from any of the neighborhood kids. They receive the free treat without the required work. In this way, both the orchid's and the children's mimic works to procure what they need with the minimum effort.

7.5 Mrs. Doubtfire at the Fight Club

Halloween is also a great time for adults to dress up and revel in the night. Favorites for adult costumes can be the latest politician and their scandals, the newest internet meme, or movie characters. This is also a time of the year that cross-dressing is fairly common as a standard costume. Whether women dress as men or men as women, this costume seems to be a yearly favorite. The cross-dressing theme occasionally appears in movies including roles that produced memorable performances for the actors involved. Glen Close in "Albert Nobbs," Tony Curtis and Jack Lemmon in "Some like it Hot," Jaye Davidson in "The Crying Game" and Gwyneth Paltrow in "Shakespeare in Love."

The late and great comedic actor Robin Williams created an endearing and hilarious cross-dressing grandmotherly persona called Mrs. Doubtfire in the movie of the same name. Within the movie, the persona, Mrs. Doubtfire, was Williams' way to stay close to his children as a nanny after a bitter and nasty divorce. During the movie, his ex-wife begins a relationship with a potential suitor which creates conflict between the ex-husband and new suitor. The cross-dressing nanny serves to minimize this conflict with the potential suitor and allow the Williams character to be involved in his children's life. In order to increase the effectiveness of his costume, Williams applies some prosthetics, lots of makeup, and to top it off, a feminine perfume. The cryptic costume allows William's character to hide in plain sight without having to clash with his ex-wife or her new romantic interest. Nature too has its share of cross-dressers and the use of nature's perfume is necessary to pull off this deception.

A cross-dressing perfume is exactly what a certain group of beetles has evolved. Insects are the largest group of animals on the plant and most certainly have the most biomass among the animals on earth. Within insects, beetles are the most numerous types of insects, and some estimates place beetles at 25 % of all known types of animals. This large abundance of beetles have inhabited all terrestrial habitats except Antarctica. This enormous range of habitats has resulted in a wide variety of lifestyles for beetles. Among the beetles, *Aleochara curtula* (a member of the Rove beetles) has evolved an interesting life stage that involves the chemical crypsis that allows juvenile beetles to hide in plain sight. The work of Dr. Klaus Peschke has illuminated this insect version of Mrs. Doubtfire.

Aleochara curtula spends its entire life feeding on the larvae and adult forms of the dipteran (flies) blow fly. The blow fly itself feeds on the dead carcasses of a large diverse set of animals. Carrion as a food source is a somewhat rare and prized delicacy in nature. Carrion is prized because once an animal dies, there are a host of organisms that will descend upon the dead animal to consume any and everything available. Needless to say, carrion doesn't last long in nature, so animals that are slow to respond or find this food miss out. Those animals that do find a good meal just lying around defend it with a great deal of intensity. Even though the *curtula* beetle doesn't actually feed on the carrion directly, the carrion is a necessary breeding ground for its food, the blow fly.

Consequently, this member of the Rove beetle is found near the carrion being fed upon by blow flies. The largest adult male beetles will patrol and protect this resource with intense aggression like some muscular bouncers outside the hottest night club. Just like the bouncers refusing entry to the "wrong" patrons, other smaller males in the vicinity are either chased off or actively attacked by the males. These fights can escalate to fairly intense levels which can result in the biting of antennae and tearing of limbs. The beetle fight club continues night and day as the dominant males guard access to their food source while the blow flies grow and develop.

Females, on the other hand, walk right by the bouncer males and which give them complete and free access to the blow fly buffet inside the carrion. The females have essentially unfettered access to all of the blow flies they desire. So, while the big bad males fight for supreme dominance, females wander in and out to eat at their leisure. Instead of a bouncer's list of the "in" people to allow into the night club, the male beetles recognize female beetles by their characteristic perfume. The female

Fig. 7.3 Rove beetle night club

A. curtula beetles produce a sex pheromone that signals to the males that they are indeed female and may be ready to mate. This odor is the free pass that allows the beetles behind the velvet rope manned by the beetles (Fig. 7.3).

If this was the whole story, the smaller and juvenile males are stuck outside looking in at two types of resources: food necessary to get bigger and stronger and females necessary to reproduce. Here is where evolution in the form of Mrs. Doubtfire provides the perfect solution to the juvenile male problem. All juvenile males also produce the female sex pheromone. Their cross-dressing costume is simply the right perfume, and Dr. Peschke has shown through chemical analysis that the juvenile odor is identical to the female sex pheromone. Juvenile's "wearing" this chemical costume are treated just like females and are allowed access to the blow flies and the carrion. Mating and feeding for this beetle occur at the same place so on the perimeter of the carrion is where the fight club exists and where males limit access by aggressively defending their territory. On the carrion itself is the singles night club where food and sex occur. The juvenile males sneak into the night club in their feminine odorous disguise in order to feed and survive. Occasionally, the juveniles have to put up with the sexual advances of the older males as their sex pheromone initiates copulation in the males, but the occasional homosexual copulation is inconsequential to access to food. Once the smaller beetles gain enough mass to be competitive during the fights, their bodies stop producing the female sex pheromone. The cross-dressing costume is removed, and they adopt the male role of fighting around the perimeter of food sources.

The large males, after spending significant amounts of time fighting other males and copulating with females, have depleted energy stores. Spending all of their time at the fight and night club takes a lot out of the male beetles. Thus, long-lived beetles need to replenish their energy for another round of activity. After another series of interesting experiments, Dr. Peschke found that long-lived adult males can still produce the female pheromone. Juvenile and smaller males simply don't lose the ability to produce the sex pheromone. Turns out that males can control the production of the sex pheromone. If a male needs to feed and not fight, that male produces the aromatic disguise and is treated like a female. These tired males then feed at their leisure to regain their strength. When ready, they again remove their costume. Other males begin to fight with them and females are attracted to them.

During several scenes in the Mrs. Doubtfire movie (and in several other comedic cross-dressing movies), the main protagonist is often caught where both the original character and the cryptically cross-dressed character need to be in the same place at the same time. So, the comedic part for the audience is observing the quick change abilities of the protagonist and wondering if the character can keep the right persona straight. For these beetles, the quick change between being male and being a cryptic female is simply turning on or off the right perfume at the right time. A quick dab of Curtula no. 5 and a drag queen is born.

7.6 The Bait and Switch Model

Charlie Brown didn't just have costume troubles during the Halloween season. Here in the States, Halloween falls right in the middle of the passionate football season. While kids may be planning pirate, ghost, or Elsa costumes for the big day, sports fans are planning parties for the big kick off of America's favorite sporting season. Charlie Brown must be a football fan as demonstrated by one of his lifetime goals: kicking a football being held by Lucy. Anyone familiar with the comic knows the outcome. As Charlie Brown gets a running start, he thinks over and over that this is the time. This is when he actually kicks the ball and watches it sail off into the sky. Yet, every single time, right when Charlie Brown gets ready to kick the ball, Lucy pulls the ball away and poor Charlie Brown ends up on his back wondering why he fell for Lucy's bait and switch again. Despite a 100 % failure rate on kicking the ball, Charlie Brown still gets lured back into the game with what appears to be sincere promises by Lucy to hold the ball in place. The payoff for Charlie Brown, kicking the football, is too powerful not to take the risk of falling flat on his back.

Animals, too, are faced with similar decisions in that payoffs need to be compared to risks. Some organisms actually have the cognitive skills, maybe even better than Charlie Brown, to effectively determine the cost-to-benefit ratio of certain behavior choices in order to maximize their rewards. The Red Knot, a shorebird, measures the benefit of consuming different mollusks against the time and energy needed to open their shells. Bees, shrews, and a host of other birds have been shown to perform risk sensitive foraging. Within this context, the foragers assess some level of predation risk and determine whether a food location or item is worth their foraging effort. I often perform a cost-to-benefit analysis when the dessert course arrives when eating out. Is that cheesecake really worth the extra money and calories?

Like Charlie Brown, I too often take the risk (eating the cheesecake) and have to pay the extra cost of more running or cycling later.

In some instances, the benefit is just too good to pass up, as in my cheesecake decision or the late night purchase of a miracle kitchen knife being hawked by some person yelling out of the TV. These benefits seem to tap into some undeniable need within us. The cheesecake delivers that creamy sweetness that stimulates the taste buds in the mouth which, in turn, sends messages to the pleasure centers of the brain. The sugary chemical signal can override the rational part of the brain that may be telling me that I really don't need the calories. Since the neural pathways that deliver chemical information in the human brain tend to bypass the cognitive areas and go straight to the pleasure centers, these cost-to-benefit analyses of behavior can be skewed toward the benefit, while not fully measuring the risk.

As mentioned several times in other chapters, the role of chemical signals in the insect world really can't be overstated. In particular, sex pheromones (as the next chapter details) are powerful signals that provoke predictable behavior within insects. The classic example of sex pheromones is of the male moth flying miles and miles upwind to locate females in order to mate. The attraction caused by these signals are akin to Lucy holding the football for Charlie Brown. He just can't resist making another attempt to kick that ball. Although with sex pheromone, there is no real risk unless there are bolas spiders around.

The bolas spider is part of a group of spiders called orb weavers. Orb weaving spiders build the classic circular web found in *Charlotte's Web* or in real or fake haunted houses. The nets are filled with sticky web strands that trap flying insects which provide food for the net spinning spider. Bolas spiders are unique among the orb spiders in that they don't spin a web; they construct a sticky blob on the end of a long thread. Their name, bola, comes from the similarity in appearance between their weapon and the weapon made famous by the Argentinian gauchos. The resemblance goes beyond the appearance. To hunt, the bolas spiders typically string a single strand of silk between two holdfasts (leaves or sticks) and hang upside down from this guide wire. While hanging upside down, the spiders begin to twirl their bola in ever increasing circles below their head and actually look like upside down gauchos. Bolas spiders feed almost exclusively on male moths. When a male moth approaches within striking distance, the spider throws the bola at the moth and if lucky, the bola sticks to the moth, an ingenious method of hunting and a different use of their silk than their other orb weaving brethren (Fig. 7.4).

Hanging upside down and twirling a bola until a random moth appears is definitively a unique method of hunting, but seems to be quite a waste of energy. The likelihood that a moth will come within the range of the bola is rather small, so the bolas spider does something quite interesting to increase the odds of a male moth approaching within striking distance. Just like Lucy uses a football and the promise of a sweet reward (finally kicking the ball), the bolas spider entices male moths with the old bait and switch method. Bolas spiders have evolved the ability to replicate and produce the sex pheromone of the female moth. Even more amazing is that each different species of bolas spiders focuses its predatory bait and switch methods on 1–3 different moth species. The bolas spiders, as far as spiders go, are fairly unat-

Fig. 7.4 Bolas spider on horse with bola

tractive spiders. Their abdomens appear enlarged and almost as if they have a flat snail shell on their back. The work of Dr. James Tumlinson (Penn State University) has worked out the chemistry of the fake male sex pheromone. In the abdomen of the bolas spiders are special glands that secrete a suite of chemicals that are identical to the sex pheromone of different moths. As stated above, different species of bolas spiders have different chemical secretions that match the pheromones of different moths. Once the bolas spider has built the sticky bolas, the spider places on the end of the bolas its mimicked sex pheromone and begins to swing the bolas in the air. Male moths in the vicinity readily take flight and are probably motivated to fly toward the scent because they think there is a female moth ready to mate. Unfortunately, the scent is a nasty deception and false advertising. The only thing that awaits the moth is a swift capture by the spider.

The potential payoff for the moth, in responding to the pheromone, is a mating opportunity and from an evolutionary point of view, mating is the highest reward possible. Thus, the selective pressure to respond to their sex pheromone is too much for the moth to pass up and the bolas spider trick works. Unlike my cheesecake choice for dessert or Charlie Brown's decision to run up to the football being held by Lucy, the moth really has no evolutionary choice. Most of the time that a sex pheromone is smelled, the reward at the end of the flight is mating and reproducing. This positive feedback mechanism, fly and then mate, keeps the moth responding to the sex pheromone. The bolas spider has hi-jacked this reproductive need and signal using a Lucyesque bait and switch strategy through copying the chemical signal that triggers the moth's behavior. Charlie Brown continues and will continue to take that long run up to the football over and over again. Each time thinking that he will finally get his reward and kick the ball only to land square on his back. Nature's equivalent to Charlie Brown, the moth, will continue to fly towards its Lucy (the bolas spider with the sexy smell) only to find failure at the end of the sticky bolas.

7.7 Removing the Odorous Mask

In the examples above, true intentions were all hidden from the perception of others through the use of masks. The masks could be visual as in the Halloween costumes, the bee orchid that has petals that resemble the female, or Lucy's promise to hold the ball for Charlie Brown. The deception could be performed with words as the politicians portrayed in *House of Cards* or, importantly for this book, chemicals as in the pirate perch or a Rove beetle. The commonality in all of these situations is the need to mask the signal or at least the intention or message of the signal being produced. We, as a species, are most likely the champions of deception which becomes readily apparent when our daily rituals are considered.

As a field biologist that works in the muck of freshwater systems, I probably put less effort into my morning routine that creates a "look" than most other people. I have never been one to care a lot about my personal appearance, but there is certainly a minimum standard that is probably expected from a societal point of view. My morning preparation ritual consists, in no particular order, of brushing my teeth, shaving, showering, and dressing. When I brush my teeth, the toothpaste usually has a minty or cinnamon flavor to it to conceal any potential offensive bad breath. While there is most certainly a dental benefit from brushing, the addition of a flavoring only serves a societal preference to have a certain set of less offensive odors from my mouth. I shave my neck and cheeks as I currently wear a goatee. My shaving cream is also infused with odors and even a chemical that stimulates the trigeminal nerves that are along my cheeks. The menthol in the shaving cream gives me a cooling sensation, and the fragrance has a floral scent which has been given an interesting name by the company's marketing team. During my shower, I use a shampoo which has its own distinctive perfume which I follow with a soap that also has a separate odor. After I dry off, I apply a deodorant as well as a body spray to add two more perfumes. Finally, to top off the bouquet of personal perfumes, my clothes have been washed in a detergent that is designed to leave clothes smelling like a spring rain. In the span of 15 minutes, I have gone from pure and unaltered body odor to at least six different applications of aromatic chemicals all designed to mask or hide the underlying personal odor. None of these applications have any health or other benefit; their sole purpose is to deceive other's noses and perhaps to please your own nose.

I wear these every day. These are my olfactory versions of the pirate and ghost costumes that visit my house during the Halloween season. The Halloween costumes allow the kids to partake in the yearly adventure to fantasize and receive candy, whereas my chemical masks allow me to walk around crowded and public domains without drawing too much attention to myself. I would imagine that not showering after a particularly arduous run or cycle might make my personal interactions shorter. I think those around me would begin to give me wide berths or begin to call me rather than talk in person. As attractive as this might appear to introverts like me, there are still limits to what I could get away with and still be employed as a professor. A quick survey of the personal hygiene industry seems to indicate that my morning ritual is not unusual as this particular business sector is a multibillion dollar industry.

Not all of this primping and preening is done to mask the deceptions of the person using these items. Yet, the sheer amount of money and volume of time spent on our need to cover up or alter the body's natural odors says something about our psychological need to control how others perceive us. Recall the imaginary elevator ride described in the previous chapter (Chap. 6). The one where you are on your way to a meeting and a gentleman enters the elevator with you. Your judgments on this person are strongly colored by what you smell. The cologne, soap, clothing detergent, and shampoo shape your conclusion on the attractiveness or competence of this person even without knowing a single thing about him. Although unknown to most of us, our morning aroma rituals help determine how society "sees" us or judges us.

7.8 The Beautiful Smell

"Beauty is only skin deep" is the oft-quoted saying, hopefully, cajoling the listener to look beyond the skin to a deeper sense of beauty. Certainly the amount of money that our society spends on plastic surgery, health fads, and fashion seems to indicate that appearing beautiful is a critical aspect of being a social animal. Despite the obvious and ironic facts that sun tanning is unhealthy, our society tends to associate a tanned and toned body with health and beauty. What if beauty is not really skin deep, but is really "nose deep"? Is it possible to alter the perception of beauty by applying the right perfume? Does Chanel No. 5 make one more attractive to other people than the bargain brand perfume? Can odors and perfumes sneak their way into our brains and influence the decision we make about those people around us?

Our cognitive brain makes judgments on beauty in our partners, and some interesting work on the visual system and beauty has identified key elements in facial features that are associated with attractiveness. Facial symmetry is a top element that is important for our brains to see someone as attractive. The troubling aspect of all of this work is that the nose has been ignored. Given the almost direct connection between our nose and our emotional center, the smell of beauty may be more important than the vision of beauty. The interesting answer to the above questions is yes. Our perception of the beauty of others is secretly swayed by what we sense with our nose. The power of odors to alter or mask the perception of reality for us has been demonstrated very nicely by the work of Drs. Johan Lundström and Janina Subert at the Monell Chemical Senses Center.

Imagine being asked to sit at a desk and watch a series of women's images appear on a screen in front of you. Each image is a straightforward picture where the women are attempting to show no emotion at all. The women are doing their very best poker face as the picture was taken. As each image appears, you have been instructed to judge the attractiveness of the women in the picture. As the tests go on, you may notice a faint smell in the background. At one point, the odor may smell like roses; at other times, there is a faint aroma of fish in the room. Still, you continue to focus and rate each of the images. The experiment described above is a paraphrase of what the Monell researchers did. The authors selected women around the age of 25 to sit and rate the attractiveness of a series of photographs of women around the age of 42. During the rating periods, the scientists delivered different odors to the noses of the raters. The odors were strong enough to be noticed but not strong enough to interrupt the task at hand. The two odors, rose and fish, were commonly known odors and differed greatly in their level of pleasantness. Interestingly, when being stimulated by the rose odor, the women rated the photographs as more attractive. Conversely, when stimulated by the fish odors, the exact same photographs were judged to be less attractive.

The odors being delivered weren't directly connected with the photographs such as the whiff of perfume that is sensed when shaking a person's hand or entering an elevator with them. Still, the odors subtly change how the photographs were perceived and in particular, whether the photographs were attractive or not. Certainly, findings such as these give some credence to our morning rituals as we prepare to interact with other people. If this research trend continues, maybe there are odors that could project other positive aspects. Maybe science could develop odor masks that serve the same or better function that all of those Halloween costumes serve. One perfume could help someone exude confidence while another one could project sincerity. At this point in time, these odor masks are still just as fake as Charlie Brown's ghost sheet.

7.9 Hiding Amongst Ribbons of Odors

In the Midwest of the United States, apart from football and Halloween, fall also brings a ritual so dear that schools allow excused absences to partake in this event. The event? The start of deer hunting season. Hunters wait in anticipation for this annual event and for a good percentage of the hunters, I think the camaraderie of tramping through the woods with family and friends is just as important as the actual hunt. Yet, the prospect of spending time in the woods searching for a buck is enough to get scores of people prepping their gear, including camouflage. So important is the visual camouflage for hunting that several companies have made their fortune by generating just the right combination of leaves, twigs, and colors that will allow the hunter to visually vanish into the background of the forest.

Hunters know that being visually hidden is not enough for deer hunting. Deer have evolved sophisticated mechanisms to detect movement and threats within their habitats. One of those detection methods is the scent of upwind predators, including the fall hunters. The odors of predators is often the first clue that deer receive that something is amiss. These chemical cues can travel a far greater distance in the woods than any visual cue. Good hunters, knowing this, do three things to minimize their presence to deer. First, they attempt to locate themselves above the deer in canopy stands. These stands are elevated either by a raised platform or by direct attachment to a tree. Second, the hunters try to remain downwind of their prey. Third, similar to the pirate perch, they spray on chemicals designed to mask the human odor. All of these mechanisms serve to change the location or quality of their plume of human scent that gently floats downwind.

Yet, if the hunter remains on the ground, their personal scent is always upwind of some animal. Simply trying to remain downwind of every animal in the forest is an unattainable proposition for any hunter. The only way to truly eliminate the human smell is to mask their odor similar to what the pirate perch appears to have done. The hunter needs to become an olfactory ghost in order to blend into the smells of the forest as easily as the visual camouflage on their clothes.

For the nonhunter, this same ninja problem exists but in a different context. As stated previously, I spend my summers in the upper part of the Lower Peninsula of Michigan. The biological station where I perform my research owns approximately 10,000 acres attached to two lakes, a couple of rivers, wetlands, and bogs. This place is a haven to those interested in field ecology as students and researchers from all around come for the excellent facilities. Unfortunately, the same habitats that draw in students and researchers are havens for mosquitos, black flies, and biting midges (called no-see-ums because they are too small to notice until they have attacked). These animals are the hunters and we are the prey emitting wondrous signals of CO_2 and heat that are beacons of free meals. Although the roles are reversed when we consider deer or mosquitos, the problem is the same, how does one stay chemically hidden in a world full of chemical signals?

Like the pirate perch, we have the ability to hide our chemical signals by applying additional chemicals that serve to mask our human odors. Although we don't produce these masking chemicals by our own physiology, the chemical masks work in a similar manner. For mosquitos, essential oils along with DEET appear to be the best working products. The exact mechanism of action is unknown, but these aromas make us chemically unattractive to these pesky bugs. For hunters, there are all sorts of choices, some of which are based in science, while others are based on good marketing schemes. Included among the many different products are detergents that will break down odor molecules trapped in clothes and mechanical devices with fans that emit concentrated ozone that will react with the organic molecules that make up the human scent and break those molecules apart. Although the science behind this technology is not solid, these devices should work in theory. With both the masking chemicals and electronic odor eliminator, our personal odor plumes, those ribbons, and puffs of personal aromas are mixed, hidden, or removed from the aromascape of nature.

7.10 That New Car Smell

I started this chapter on lying and deception with references to politicians and the multiple levels of duplicity that seems to exist throughout politics. I find myself often watching the speeches of politicians with a sort of macabre fascination.

Knowing that lies are the common denominator in all political speeches, I become quite fascinated in the ability of these individuals to use their words to convince their audience of their point. Whether they want voters to trust, love, respect, or donate to them, the word choice, the pauses, and the hand gestures are wonderfully developed to the point that the total package of the speech is a performance art piece. I feel that the really successful politicians certainly out perform any stage hypnotist or Svengali.

These speeches work primarily through two distinct sensory channels: those of our visual and auditory systems. Like a broken record at this point in the book, this is the main focus for the transfer of important information for most of our society. As much money as the fragrance and flavor industries make, they pale in comparison to industries that focus on our visual and auditory abilities. This fact is why the politicians work on hand gestures, tie colors, and their facial movements. To put a somewhat sinister spin on these speeches, the main point is to control the mind and choices of the voters. Essentially, these politicians want the listeners to be convinced, by any means, that their particular politician is correct and deserving of their vote. Words in these cases are inherently powerful and can be used to sway the minds of listeners. Odors may be just as powerful though. Tapping into the emotional center of the brain, certain aromas may subtly sway the mind even more than any great speech through a mechanism called priming.

Within the psychological literature (and the area of magic and mentalism) is the concept of priming. Priming occurs when certain environmental stimuli or conditions influence the behavioral responses to another stimulus. Priming occurs when you head to a movie, say Mrs. Doubtfire, and one of your friends who has seen the movie before, extols the funny nature of the movie. You may be primed to think that the movie is funnier than if your friend panned the movie. Magicians and mentalists are exceptionally adept at priming audience members to think or see things a certain way. Any card trick where you select a card through a series of choices needs a hearty dose of priming to work. At the end of the trick, the magician has planted within your mind the concept that you have freely picked the card that magically matched what was revealed by the magician. In his excellent book *Thinking Fast and Slow*, the Nobel Prizing winning psychologist, Daniel Kahneman, provides a host of amazing examples of priming. My favorite example includes a series of experiments where reading lists of words can alter behavioral choices. Participants handed words commonly associated with old age tended to walk slower down a hallway than participants handed random words. In another example, real estate agents overestimated the price of houses after hearing words associated with really large numbers and another set of agents underestimate the prices of houses after hearing words linked with really small numbers.

Retailers, just like politicians, want to convince people to do something. Rather than voting, retailers want to convince the consumer that their product is high quality and therefore should be purchased. A host of psychological literature has shown that people are more likely to purchase products when they are in a favorable mood and are more likely to rate a certain product as high quality when happy. The emotional happiness is critical to keep shoppers shopping and to finally seal the deal.

Since odors have a beeline to the emotional seat of our brain, aromas may be a key priming device to influence people's shopping behavior. Interestingly, placing chocolate scents within a bookstore increased the amount of time that shoppers studied books and also increased the purchase of books by 40 %. Vanilla scent candles reduce stress and when placed within a shopping environment caused shoppers to linger over products.

The ability of odors to influence our shopping behavior goes beyond that of purchasing. In another study, researchers showed that people walking through a mall were more likely to help people of the same sex if there were pleasant odors present. In this study, the accomplices (those participating in the study) either dropped a pen or asked for change for a dollar bill. When surrounded by odors of baked cookies, coffee, or pastries, people were more likely to either help pick up the pen or provide the change requested. If the same task was performed without the ambient pleasant odors, the percentage of people helping dropped by half. In these cases, the fragrances were altering the behavior of the participants. In many ways, these odors were priming the individuals for a certain set of behaviors just like politicians, comedians, and magicians prime individuals with their words and actions. Just like that new car smell that envelopes the nose during an attempt to buy a new vehicle, these odors sneak their way into the brain to provide a gentle or not-too gentle push to buy a book, help a stranger, or potentially, vote a certain way. I use the term sneak because we are not cognitively aware that these odors are influencing our behavior. Whether intentional or not, our minds can be deceived to believe or do things due to the deceptively influential nature of chemical signals.

Chapter 8
The Allure of Sex

Fifty Shades of Odors

During the current semester, I am teaching the ecology section of introductory biology. This class is often populated by biology majors, related majors such as environmental science, and those students needing to fulfill a university science requirement. These classes are rather large, typically filling to 140 or more students. The largest classroom in the biology building only sits 100 students, so I have a 10–15 minute walk across campus to a larger classroom. In redesigning the course to hopefully maximize student's learning, I am teaching this large lecture as a small discussion class, which requires the help of my graduate students. They are also involved in running the laboratory section of the course, so their presence in my lectures helps to integrate the two parts of the course.

On this early September day, the weather is perfect. Teaching at 9:30 am, the midmorning temperature is already at 72 degrees and a clear day bodes even warmer weather. This walk is turning into an enjoyable stroll and since we are in a rather jovial mood, the short conversation we have is full of humor and laughter. All of my students have some sort of behavioral aspect to their research project, so in addition to our jokes, we are all engaged in putting our observational skills to work. Our walk takes us straight through the heart of campus, and the walkway is full of students heading to all corners of the university. On typical days, a significant portion of the walkers have their heads down playing with their phones. Although not as dangerous as distracted driving, distracted and unaware students walking with their heads down can turn our walk into an interesting game of "dodge the student." While modulating our direction and speed to miss students, we use our enhanced perception to watch other interactions occurring around us. Who will be nice and allow someone to cross in front of them? Can we tell the nature of a conversation by the hand gestures of the two speakers? Which students are in a hurry and seem flustered and which ones are out for a relaxing stroll? I wonder if we were as acutely aware of the chemical environment, we would enhance our conclusions about our observations.

After a rousing class period, I am energized as are the graduate students. Amid our animated conversation in the class, one of my students suggests that we stop and get a coffee at the student union. Teaching class is part intellectual and part emo-

P.A. Moore, *The Hidden Power of Smell*, DOI 10.1007/978-3-319-15651-4_8

tional. Feeling a little drained on the emotional front, I head with the group toward the coffee shop. The union entrance leads into a large open area where groups of faculty, staff, and students gather for meetings, conversations, and eating a meal. We have entered mid-morning, so the lunch crowd has yet to really fill the space. As we make our way over to the coffee shop, I really begin to study the different groups of students scattered around the tables. I can see a group of three students staring with some level of intensity at their computer screens. Around them on the table are different papers and a couple of textbooks. I imagine that they are working on a project for a class, and as I pass their table, I spy equations in their textbook. Math or physics is probably a good guess on the subject area.

Two tables down from this group are two older gentlemen dressed in suits. Both have large coffees in their hands as they stare at a large set of physical property plans. The campus is currently going through a building phase and I guess that they are discussing some question about what gets built next. The coffee shop has its own tables and some rather comfortable chairs in the far corner of the shop. As we walk into the shop, I immediately notice two students sitting in the corner casually drinking coffee and staring out of the window onto the lawn of the campus. No books are open and no conversation is taking place. At least, no auditory conversation is occurring. Every once in a while, they'll exchange glances, and one of them will reach out to touch the hand or arm of the other.

Most of the individuals on a college campus range between the ages of 18–25 (if the graduate students are included). For many individuals in this age group, the social dynamics of their groups center around two fundamental elements of maturity. First, many of these students are continuing to discover who they are as far as their value systems and their personas. Having made the transition from the awkward period in high school, there is a chance to reinvent yourself by shedding the labels that were attached to them during their previous schooling. Second, these students are in that stage of life where they are seeking someone that could be their partner for the rest of their lives. The pair in the corner of the coffee shop seem to be part of this second group.

For this pair of students, something has brought them together. Maybe a shared enjoyment of a certain type of literature or music or a series of classes that allowed them to study together. Whatever started them on this journey, their attraction is reinforced by the closeness that envelopes them. Certainly, the visual nature of attractiveness has been studied over and over among human males and females, and the current consensus on beauty seems to center on symmetry of facial features. Secondary features, a partner's touch, the softness of their voice, a small smile, also seem to reinforce the connectedness that they feel toward their romantic companion. As the pair of students rise to leave the coffee shop, their hands instinctively reach out to grab the other and they leave hand-in-hand.

One of the most fundamental functions that organisms perform is reproduction. So important is this singular function that Charles Darwin wrote a separate companion book to "*On the Origin of Species*" that focused specifically on the selection of partners for mating among animals called "*The Descent of Man and Selection in*

Relation to Sex." Although still a part of natural selection, Darwin thought that the sexual selection process could run counter to natural selection under the right conditions. In these cases, sexual selection could favor traits that appear to inhibit the ability of the organism to survive. Two classic examples of these cases include the tail feathers of the showy peacock and the exquisitely large antlers of the extinct Irish Elk. These features are present and are favored by natural selection because the female peahen and Irish Elk are attracted to these traits for reproductive partners. These features also have detrimental effects as the peacock can no longer fly and the Elk's antlers inhibited its movement, but the benefit of increased sexual attractiveness outweighs these costs.

The majority of life on this planet reproduces asexually. The lack of any need for a sexual partner occurs in bacteria, fungi, sea anemones, corals, and plants such as strawberries and potatoes. These organisms can reproduce through fission or splitting themselves in two: budding or by a handful of other mechanisms. For the rest of life, there are three central problems of sexual reproduction. First, potential mates are often dispersed across a wide territory, so the need to locate a mate is a dilemma. Second, even if a potential mate is located, they may not be sexually mature and ready to mate. Organisms develop sexually through time, and there are periods of time when organisms are reproductively ready and other times that they are not. In a number of species, the females and/or males cycle through sexual receptivity. Locating a mate before sexual maturity or when they are not in the right part of the cycle is a waste of time and energy. Finally, even if a mate is found and is sexually ready, the quality of mate needs to be assessed. This assessment is the sexual selection that Darwin wrote about in his second book. The choice of a mate (or sexual selection) is crucial because unlike asexual reproduction where the offspring are identical in genetic makeup, sexual mating results in a mixing of the gene pool of two organisms. If the mate choice is poor, then the offspring of such a mating could be less attractive or less viable. Advertising sexual readiness or receptivity and the quality of ones genes is at the heart of reproduction.

Although a wide variety of organisms use visual or auditory signals to inform potential partners of sexual awakening, this area of biology is dominated by chemical signals. The field of chemical ecology rose to prominence through the understanding of sex pheromones in insects. As detailed in Chap. 2, identification of the sex pheromone bombykol used by silkworms launched a field that still hunts for different chemicals used in different aspects of reproduction. Since that discovery in 1959, the field of chemical ecology has unveiled chemicals that animals can use to determine whether potential mates are in a reproductive state or of the appropriate quality in order to mate. On the wilder side, other chemicals can make sexual slaves, start mating frenzies, castrate individuals, and put others in chastity. This range of uses of chemical signals for sex is certainly more than 50 shades of chemicals. The complexity and nuances that these signals are being used for represent those shades, and some of these shades would make the Marquis de Sade blush.

8.1 The Siren's Song

For that pair of students in the coffee shop (and most of the other students on campus), the sexual call of nature (through hormones) mixed with social expectations drives the search for a companion. Current debate within the scientific community is exploring the relationship between evolutionary forces (internal physiological mechanism) and societal pressure to find a partner. Our heightened view of our own intellect coupled with our insistence that our society is independent of nature's forces seems to create the concept that social pressure to find mates is far more important than anything that nature throws at us. Yet, research is mounting that demonstrates that the hormonal level of our physiology is a powerful driving force for behavioral choices. A college campus is an ideal place for this research given the age of most of the members of the student body and the close proximity of all of the students. The attraction, courting, flirting, or other activities are almost guaranteed. I do not mean to insinuate that the sexual drive and activity of young adults is mindless, but some acknowledgement of coupling of certain stimuli with the hormonal levels present at this age can guide behavioral decisions such as seeking out a mate.

In Greek mythology, Odysseus ran into the ancient equivalent of nature's sexual call while sailing past the famous and deadly sirens. The sirens were a group of beautiful sisters located on an island surrounded by rocky shores. Some poets and authors have named them the muses of the lower world, referring to the irresistible nature of their singing. The song was supposed to wrap the poor listener in complete and total rapture. The mind of any sailor who was unlucky enough to pass too closely would be a slave to the song of the sirens. Sailors would then be drawn into the rocky shores to their deaths.

Odysseus was quite aware of the sirens' power, so he plugged the ears of his crew with bee's wax to ensure that they continued to power their boat forward past the deadly sirens. His sailors bound him tightly to the mast to ensure that he was not drawn to the deadly islands. Although the song enticed him greatly, Odysseus was bound so securely he could not escape. Safely passing the siren's song, the crew continued their journey home. Transcending Greek mythology, we find sirens throughout the real world of animals. For some animals, the call of their sexual "siren's song" is not so easily ignored, and unfortunately, following the chemical song leads to their death.

The dark fishing spider (*Dolomedes fimbriatus*) is a rather large spider (body can be 1.5 centimeters in length) found near both forested and aquatic habitats. Within these habitats, the spiders are usually found on structures near water like logs or docks. Fishing spiders are voracious predators and will consume large amounts of aquatic insects. They have special adaptations on their front legs that allow them to detect the vibrations of aquatic insects either on the water's surface or even below the surface. In essence, they "hear" the movement of their prey like larval dragonflies, stoneflies, and other aquatic insects. These spiders even have special hairs on their body that are unwettable, which means that these hairs are so hydrophobic that they literally cannot get wet. These hairs allow the spiders to run across the water or

Fig. 8.1 Spider as siren

even take short dives to capture their prey. During their scurries across the water, they often attach a dragline to a solid surface that helps them return to their dock or log island of safety (Fig. 8.1).

Just like Odysseus' sirens, the females will locate themselves on a preferred habitat during mating season. Once they have located their own special "island," they begin to sing their chemical song. A sex pheromone is rather typical for insects and spiders and helps coordinate the timing of mass matings (as in those clouds of insects so often seen during the summer time) or to lure potential mates from great distances. The sex pheromone is carefully placed on the dragline, which increases the surface area of exposure for the pheromone. The surface area is important because of how odors are carried by the wind (See Chap. 2). Just like a siren's song, tendrils of odors waft downwind and will lure in the lustful males. The difference between the mythological sirens and the dark fishing spider is that the male spiders actually get rewarded for responding to the sex pheromone. Odysseus and his crew would have met their fate on the rocks. The male fishing spider makes it to the female spider. So the promise of sexual intercourse is not an illusion as with the sirens, but reality for the male spiders. The sex pheromone, as with most pheromones, is an alluring mix for the males. So powerful is this call, that males from all around are drawn to the female.

As the males approach the females, there is a sort of courting dance that occurs. The dance consists of a series of drumming the ground with different legs to produce the right vibrations. If the vibrations are the correct intensity or rhythm, then the female allows the male to copulate. Unfortunately for the male, their odyssey is a one-way journey and they'll never return home again. After copulation the female kills and eats the male in an act of sexual cannibalism, an act that is not all that rare within insect and arthropod societies. If the female happens to not consume the male, the sexual act itself actually mutilates the male spider which eventually dies after consummating their final act.

Consuming the male provides the female with additional energy for egg production. The female gets an evolutionary advantage through this increase in number of eggs. The male contributes to the health of its offspring by giving his body up for nutrition for the female, and the advantage for the male is the additional eggs carrying his genes forward. In fishing spiders, the males are obviously monogamous meaning that they mate with only a single female. The female, though, does not follow the same strategy because a single male does not satisfy her reproductive needs. The female continues this pheromonal singing over and over again for multiple males. The males, like the sailors of old, are drawn to the female by the alluring sex pheromone. The seductive nature of the fishing spider's perfume is nature's equivalent to the mythological Siren's song.

8.2 The Right Smell at the Right Time

Within ecological theory, the concept of niche is important for understanding where organisms are located. Although used in the vernacular now, niche can be formally defined as an environment that has all of the abiotic and biotic needs of an organism. If resources (abiotic and biotic) are plentiful, then ecologists often find that niches can become narrower and narrower. On average, organism's niches in equatorial habitats are smaller than those found in the temperate areas. This is primarily because of the stability of the equatorial habitat as opposed to the large seasonal variations found in temperate parts of the world. As with any good economic marketplace, if there is a niche where money (economic resources) exists to be taken, a company or set of companies will arise to fill that niche. In the last decade and a half, the rise of the online dating market seems to be a healthy and lucrative area. Money is there to be taken. If enough money is plentiful, then, just like in ecology, niches (online dating market) can be narrow.

Within the online dating world, niche specialization seems to be occurring. After a couple of larger companies demonstrated that the marketplace of online dating was viable (like Match.com or eHarmony), specialized companies began to evolve. Instead of aiming or broadcasting their advertising to a large audience, these other companies have targeted specific groups. Companies like ChristianMingle.com and Jdate.com are aimed at specific religious groups, while others like Tastebuds.fm (matches based on music tastes), AnastasiaDate (connects Russian women with Western men), and DateMySchool (dating those that are alumni or current members of a specific University) aim their services to a much smaller and more focused audience. Communications and signals can be tailored with the smaller niches. For example, bringing people together using music tastes (Tastebuds.fm) is a significantly different method than bringing people together because they are Jewish (Jdate.com) or Christian (ChristianMingle.com). Using the wrong type of communication will lead to failure or at least failure to attract the right clientele or right mate. Animals using chemical signals are faced with the same problem as humans using online dating services. Attracting a mate is only one aspect of the problem that is solved with pheromones, as animals need to attract the right mate. A silk moth that releases a sex pheromone that draws in a paper wasp won't be that successful in the mating department.

Most of the other chemical signals touched upon in this book are broadly tuned signals. Within the world of sensory biology, tuning of a signal refers to the specificity of the stimulus properties of the signal. If a signal is a broadly tuned signal, then it will stimulate and cause behaviors across a wide spectrum of organisms. In the aquatic environment, feeding cues are broadly tuned. Most crustaceans will respond to organic acids like taurine, alanine, or tryptophan with a host of feeding behaviors. Even aquatic fish will respond to these acids with feeding behaviors, so these types of chemical cues are perceived to indicate food for a wide variety of organisms. Food and social signals tend to be more broadly tuned than those signals associated with the need to identify individuals or individual species.

Sexual signals, in general, and sex pheromones, in particular, are very narrowly tuned signals. Through evolutionary selection, these signals activate the senses and cause behavior only in the correct species such as the silk moth or paper wasp noted above. If the sex pheromone were to draw in a multitude of species, animals would spend additional energy in finding mates among a mass of other species or even end up mating with the wrong species. Natural selection has favored those sex signals that are aimed at one specific target. Conversely, with food cues, the selective pressure for organisms is to detect as many food sources as possible, so chemicals like taurine, which is found in almost all organisms, carries information about food that all organisms need to survive. Organisms can be classified by feeding guilds such as omnivores, carnivores, or herbivores. The commonality of food sources across these guilds is also reflected in the sensitivity of their senses to chemical signals. Herbivores respond with feeding behaviors when stimulated by a broad spectrum of plant compounds. The same can be found for other feeding guilds like carnivores. Sexual signals are specific to the species and responding to the wrong sexual signal is a waste of energy and time.

In most habitats, organisms are surrounded by closely related species. A simple walk through a temperate forest habitat would reveal hundreds of different insect species. A broadly tuned sex pheromone would draw in a crowd of lusty males or females, but the wrong crowd. Very few of those insects present in that crowd would be of any sexual benefit to the signaler. To help discern with this narrow tuning of sexual call, species have evolved pheromone blends. Parts of the chemical blend could be identical across closely related species, but as long as the total blend is different, the signal would still only stimulate a single species. Across closely related pheromone blends, single molecules could be different. For example, one of the pheromone compounds could simply contain an additional carbon or an added nitrogen somewhere along the carbon backbone. This singular addition or change could be enough to allow animals to differentiate between signals. Sometimes these structural changes to the chemical even cause different behaviors in closely related species.

The Mediterranean corn borer (*Sesamia nonagrioides*) is a small moth found in southern Europe and in some parts of northern Africa. The larvae of this moth feed on a number of different plants including corn. These species, and closely related ones, are not only tremendous crop pests but are very invasive. The larvae of the moth bore into the stem of the corn plant which effectively kills the plant. Like many of the insects outlined in this book, the moth has a sex pheromone which is actually a blend of different chemicals.

A closely related species, the European corn borer (*Ostrinia nubilalis*), overlaps in habitat with the Mediterranean corn borer. In addition, the European corn borer also feeds on similar plants as the Mediterranean corn borer, including corn. The overlap in territory and in feeding guild could create a problem when mating season rolls around. If the two species used broadly tuned sex pheromones, there would be interbreeding of these two species which would lead to failed viable egg production. Like the differently "tuned" online dating sites, using the right signal to communicate is essential to finding the right mate. Both species have pheromone blends made of three compounds each, and each of the species has three distinct compounds. This is the language of smell and the right three compounds in the right concentration differentiate between species. The word "ear" uses the same three letters as "era" and "are" but has significantly different meaning. So there is no overlap in the pheromone components. Having the right pheromone works well, as good as having the right advertisement, but the Mediterranean corn borer has an additional trick to ensure effective communication between the "right" males and females.

The Mediterranean corn borer produces a compound named Z-11-hexadecenal within its pheromone blend. This chemical attracts the male Mediterranean moths, but in addition, actually inhibits the behavioral response of the European corn borer. During the male ritual of finding the female at the end of a pheromone plume, males go through several stages of behavior. Each stage needs to be completed in order for the next stage to take place. The stages like wing fanning, take off, upwind flight, zigzag patterns, landing, and catching females are all triggered by a combination of chemical stimulation and other sensory signals. This singular compound (Z-11-hexadecaenal) produced by the Mediterranean corn borer inhibits all of these behaviors of the European corn borer. So, despite the overlap in habitat, these two moths have pheromone blends that specifically target their own species over another species effectively inhibiting the behavior of a similar species that overlaps in space. This pheromone compound is not only the right pick up line for the Mediterranean moth, but is a definite turn off for the European species (Fig. 8.2).

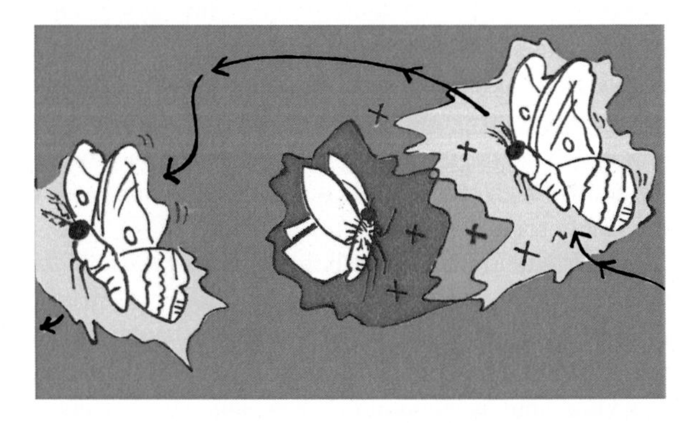

Fig. 8.2 Moths being turned off by smell

8.3 Finding the Right Dance Partner

Around the time I turned 12, about a year after puberty began to take hold, seemed to be when our school began to have dances. Unlike previous parties with music and dancing, school sanctioned dances had evolved into something different than just moving to music. The opposite sex began to have a different meaning, in addition to being a project partner or a member of a sports team. Dances seemed to become this awkward mix of attraction and repulsion. The physics of gravitational forces seemed to really be out of sync. At our dances, the girls stayed on one side of the room and the boys stayed on another side. Occasionally, some brave girl or boy would make their way across the chasm of the dance floor to quickly chat with somebody from that other group. We all wanted to be on the other side of the room talking with each other (the physics of attraction), but hesitation and embarrassment kept us plastered against our respective walls (the physics of repulsion). There always seemed to be one or two kids out dancing to any tune that was being played. Those of us against the walls really didn't think that this student was brave, just a tad bit crazy. Looking back, those students who were carelessly dancing were probably the only ones completely at peace and comfortable with who they were at that moment in time.

For my generation, these awkward dances were the beginning of our flirting, dating, or courting. All of us were thrust into the uncharted territories of sexual attraction by the hormonal changes occurring within our bodies. Probably the most awkward part was the actual dancing. Whether the song was a slower version that required some bodily contact or a faster version which featured some form of arm, leg, and torso movement, the unscripted nature of the dance itself made the entire endeavor an adventure, particularly in finding a partner that worked well with your style of dancing. Dancing with a partner that was either taller or shorter than you or one that was far more energetic with their arms and legs added to the amateur nature of our collective movements. The right dance partner would have made our movements more artistic than awkward. I distinctly remember attempting to dance with a girl that was significantly shorter than I was during a slow dance. Throughout our entire time together, I spent the totality of the dance working out exactly where to place my hands and arms so that I was comfortable, touching the appropriate body parts on my partner, and hopefully sending the right message about my level of attraction to the girl. I must have failed at one of those tasks because we never danced again after that one attempt.

None of these dances resembled anything that I had been exposed to growing up. My parents were quite fond of both musicals and period movies and the dance scenes that dominated my prepubescence were quite formal, scripted, and elegant. Since I was not sufficiently intellectually aware enough to recognize the difference between movie and real life, my thoughts centered on how these imaginary dances were always so well-choreographed and led to finding that single one person you were meant to spend your life with. When you found that right dance partner, everything fit. The music was right, your movements were in sync with their movements, and even the physics of attraction overcame those forces of repulsion. The search for that fit was similar to Cinderella where the prince is searching the entire kingdom for the single foot that was the proper size for the glass slipper. We needed the psychological equivalent of the glass slipper at those dances. If we could have had some sort of signal that would show us the one that would fit our needs those dances would have been a lot less awkward. In the natural world, the male St. Andrew's Cross spider performs his version of Prince Charming's glass slipper test, but these spiders are a lot more efficient than searching the entire animal kingdom for the right foot for the right slipper.

St. Andrew's Cross spiders (*Argiope keyserlingi*) are the typical orb weaving spiders that derive their common name from the unique construction of its web. Like most orb weaving spiders, the web is circular in shape with bands of silk traversing the circular ones at perpendicular angles. Unlike other orb weavers, this spider has four additional bands radiating outward from the middle, and these bands of silk are reinforced for either enhanced prey capture or predator protection. These enhanced bands are often quite thicker than the other elements of the web, so that the web has an X appearance to it. In addition, the spider elegantly holds itself in this same X-shape by placing pairs of legs on each arm of the X. Thus, the web and spider is said to resemble the cross upon which St. Andrew (the patron saint of Scotland) was said to be crucified and is represented on the Scottish flag.

As unique as the web of this spider is, its reproductive strategy is even more intriguing. Both male and female *A. keyselingi* spiders have paired reproductive organs. In females, this organ is called the spermatheca and in males, it is termed the pedipalp. Paired sexual organs is not uncommon in the animal kingdom, but within this group of spiders, through a series of complex morphological adaptations, male spiders can only inseminate the same side of the female and male spiders only inseminate females once. Therefore, right sided males inseminate the right side of females and vice versa. This wouldn't be such a problem for the male spiders except that after a male inseminates a female spider, part of his pedipalp breaks off to block the spermatheca on that side of the female. The males have evolved this strategy to ensure that at least half of the female's eggs contain his genes.

Within this particular species of orb weavers, both females and males prefer to maximize the diversity of sexual partners, so after a first mating, both sexes seek a second different mate. Consequently, the spider's sexual world is composed of virgin females and males with both sexual organs open and ready, along with once mated partners. If Prince Charming (a male spider with its right side sexual organ ready) tries to put the glass slipper on Cinderella's evil stepsister (a female with the left side spermatheca open), the shoe won't fit. Prince Charming will have to continue his exhaustive search for just the right dance partner. So, an efficient search mechanism is essential for these spiders.

Dr. Marie Herberstein and her lab in Australia have found just such a mechanism. Dr. Herberstein and her students placed male and female St. Andrew's Cross spiders in different test challenges. In a single middle web, they placed virgin female and male spiders; then around this central web, they placed either virgin, single mated, or double mated females in a configuration that mirrored the X shaped pattern on the spider's web. Finally, they placed small fans to create airflow from the outer females toward the central pair. Male spiders, after having mated with the central virgin female, quickly headed toward the virgin or single mated females as opposed to the double mated females. As the females construct their webs, they place sex pheromones on their guidelines of the webs, and it appears that these pheromones guide males into making appropriate and efficient choices about whom to mate with.

The male spiders were able to distinguish between single mated and double mated females based on the sense of smell. In addition, single mated males showed a clear preference for single mated females, indicating the ability to distinguish (at the olfactory level) the mating history of potential partners. Females may alter their pheromone production or even the type of signal produced as a result of being mated with once or twice. Also, the males may change their responses to these pheromones based on their mating history, but most certainly the sex pheromone and its use within this spider system communicates meaningful sexual history between potential partners. Unfortunately for these males, they were unable to distinguish which side of the female spider was available or unavailable through their sense of smell. Thus, the search for the right dance partner, those single mated females with the same side of their sexual organs open, has to continue just like Prince Charming's search for the right foot to fit the right slipper.

8.4 Getting in Tune

My mental musings about mythological Greek heroes, online dating, and awkward dances are broken by the sound of my name being called. "Doppio con panna for Paul" is enough to remind me that I am with my graduate students at the coffee shop in our student union. The group of us have our drinks and are off to finish our cross campus walk to the Biology Department. Exiting the student union, we are greeted with a steady stream of students entering the building for a mid-morning breakfast or snack. Fall on a college campus brings crisp weather, excited and energized freshman, and the blossoming color change of the campus trees. Although early in September, there are already splotches of red, yellow, orange, and brown leaves that are easily spotted on our walk back to our offices. Along our walk, we notice that a preponderance of students are also wearing their fall colors and on this distinctive day, many of the students are encased in orange and brown colors.

This particular fall Friday on our campus, and likely across any typical college campus, is the day before a home football game. The school colors just happen to match the orange and brown of the changing leaves, so there is a confluence of colors on the open sidewalks. This is enhanced by the school body wholeheartedly showing their support for the upcoming game by wearing our burnt orange and brown. To help coordinate this show of support, the school has manufactured signs and events to focus the student body on this singular event. If we were to return to the union for lunch, we would be inundated with cheerleaders and mascots cajoling the crowd into a rousing version of the fight song or the lesser known alma mater.

The coordination of events, colors, singing, and cheering is all designed to get the student body in tune with the collective show of support for one team on campus. For this purpose, the visual signals (orange and brown) and auditory cues (fight song) are meant to evoke emotional (and maybe physiological) responses such that everyone's effort, like cheering, is coordinated.

Among different animal groups, the concept of coordinated behavior is central to the theory of sociobiology. Within termite societies, coordinated group behaviors can build massive termite mounds. Workers dig, repair, and build tunnels, while soldiers protect the mound from intruders. For flying or swimming animals, swarming or flocking in a coordinated matter can help reduce the effectiveness of predators. The flurry of quickly moving fish or birds presents a difficult target for predators to single out prey. In primate societies with well-structured hierarchies, differentiation of group jobs (foraging, infant care, and defense of the troop) is central to the survivorship of the troop.

Coordinating reproductive efforts of males and females is imperative in species regardless of their level of sociality. Many animals have external fertilization which means that the eggs of the female are fertilized by the sperm from the male outside the body of either the male or the female. In fish like bass, trout, and salmon, nests are constructed by one of the two sexes. After the nest is made, there is typically a courtship ritual that involves a series of intricate swimming maneuvers, and if the dancer is attractive enough, then mating will occur. The female fish will swim past the nest and lay its eggs among the rocks and pebbles of the nest. Once this task is completed, the male swims over the nest and releases his sperm onto the waiting eggs. If the eggs are laid too early, then the eggs don't get fertilized. If the male releases his sperm early, then they float downstream and no fertilization can occur. Synchronization and timing of these acts are essential for successful mating. For those animals with internal fertilization, timing is just as critical for successful mating.

Timing of reproductive effort is important because within mammalian species, the females cycle in and out of estrus. When a female is in estrus, she is fertile from a physiological level and is behaviorally receptive to mating approaches by the male. In some primates, like baboons, the females develop obvious external signs when they are in estrus. This can include swelling and increased redness in the genitalia. The male baboons are sensitive to these signals and respond with appropriate mating behaviors. In other mammalian species, like humans and orangutans, there are few external signs of estrus and this is called concealed ovulation. When female mammals (other than humans) are not in estrus, their behavioral receptivity to mating advances is nonexistent and can often lead to aggressive responses from the female. Therefore, the males of these species are under selective pressure to either detect when females are in estrus and receptive to their mating advances or coordinate when females cycle and enter into estrus.

Of course, instead of detecting whether mice females are in estrus, males would certainly benefit if there was some way to synchronize all of the females such that they were in estrus all at the same time. Called the Whitten effect after Australian reproductive biologist, Wesley Whitten, male mouse urine has the ability to synchronize female mice estrus cycles. Among mice, the female's reproductive cycle is approximately 4–5 days in length if they are living in isolation from other females. When grouped together, female reproductive cycles are often longer in duration and become irregular in their length. If these same females are exposed to bedding that contains male urine, these females will begin to have shorter and synchronized cycles within 3 days. The male urine does need to be donated from a sexually mature individual as urine from an immature male does not alter the female's reproductive cycle. The key chemical that is involved in this effect is testosterone, which is also related to the sexual maturity of the male.

The theory behind the synchronized cycling and the Whitten effect involves the evolutionary advantage in producing offspring with the most dominant male. Male mice receive a reproductive advantage by having all of the surrounding females in tune reproductively. If a new dominant male arrives into a habitat and begins to send out testosterone laden signals, the females will become sexually receptive at the same time. The male mice will have an increased chance of reproductive success with so many sexually receptive females around. Females have a selective advantage by being able to mate with the dominant male mouse if they are in the appropriate reproductive phase when this male is around. Just like the beginning stanza of a fight song causes the student body to sing in tune, the pheromone signals from male urine reach into the brain of the female mice to bring their physiology in reproductive harmony.

8.5 A Reverse Ugly Ducking

As previously mentioned, this college aged group of individuals and their new found freedom away from high school or seniors readying themselves for their final semesters seem to drive the need for finding that someone special. Of course, special is an ambiguous definition within human culture, but within the animal kingdom special is focused on a singular issue: fertility. As I pass the many students walking to and from campus, I wonder what constitutes attractiveness for each and every one of them. Some of the men are tall while others are quite short. Probably the most variable and controllable feature for men is some form of facial hair. As I pass by these students, I see full beards, goatees, and clean shaven individuals. Throw in a smattering of very unique facial designs to create a wide variety of looks. Among the female students, hair color, length, and level of curls seem to be a tremendous source of variation, particularly when students have a host of colors in which to alter hair. I can see complete color changes such as red, blue, or green as well as streaks of singular colors added to provide some contrast. I wonder if our ability to decorate and accessorize our appearance (for both men and women) is tied to some need to appear distinct in order to attract the right mate.

Several different animals also accessorize their appearance. Decorator crabs, true to their name, glue objects to their shells in order to hide from predators. Crabs can even have other organisms like sponges and coral attached to their shells. The shells

Fig. 8.3 Bower birds and decorator crabs

of these crabs can get quite elaborate with numerous decorations and adornments. Probably the champion accessorizing animal has to be the Bower bird. In order to attract female Bower birds, the males will build a strong and sturdy ground nest. Yet, female Bower birds are very choosy individuals, and a simple nest is just not the right type of home to move in and settle down. So, the male then accessorizes the nest with anything he can get his beak on. When selecting objects, the males choose particularly colorful and bright objects, and these objects don't need to be natural. We, as a species, have provided male Bower birds a host of new shiny and colorful objects with which to decorate his nest. Our trash in the form of plastic silverware, bottle caps, and even scientific equipment has found its way into the nests of Bower birds to make them more attractive (Fig. 8.3).

As demonstrated earlier in this chapter and in Chap. 7, organisms, as well as humans, can also use chemicals to make themselves deceptively alluring to the opposite sex. Certainly, the preponderance of perfumes and scented soaps show how important our personal odor is during our social interactions. Within the animal and plant kingdoms, sex pheromones serve to send information about sexual readiness and serve to make the wearer more enticing. In many mating systems, the presence of the pheromone is a requirement for mating. In other systems, a dab of this or that chemical and suddenly an ugly duckling becomes a beautiful swan. Using the right perfume at the right time can attract a crowd, and yet sometimes a crowd is not wanted.

Damselflies are beautiful and delicate flying insects, and the females lay their eggs on the leaves that overhang aquatic habitats. Related to dragonflies, the damselflies often sport colorful bodies of red, yellow, blue, gold, and orange. In addition to this, the wings are iridescent in color. Completing the picture of damselfly romance is the fact that male and female Damselflies can form a heart shape during copulation. During particular times of the summer, small streams can be filled with colorful hearts of damselflies fluttering all around the stream. Despite this idyllic picture everything isn't as peaceful as the description above seems to indicate. As females become sexually mature, the large groups of males are drawn in for potential mating opportunities. The mating system of damselflies belies their name as the males do not act like medieval knights and the females are hardly damsels. The males begin a series of very aggressive and harassing flights and attacks upon sexually receptive females. The harassment continues until one lucky male is chosen to copulate with the female. Female damselflies can have multiple matings, so the harassment doesn't stop after the first copulation.

Harassment is costly and energetic for the female damselflies. Although exact cost calculations have not been performed, excessive harassment seems to lower the fecundity of those females. To avoid harassment, the females have evolved an alternative color morphology that resembles the male coloration. The beautiful damselfly becomes an ugly duckling as far as the males are concerned. These alternative morphs suffer less harassment and thus can avoid the energy costs associated with fighting off males. This alternative colored female also receives less copulations, so there is a fitness cost associated with this variation in appearance. Depending upon the number of males and intensity of the harassment, natural selection can favor different ratios of the traditional to alternative color frequencies. The alternative color is akin to single females wearing a fake wedding ring to avoid the harassment at overcrowded bars. The color of the damselfly or the ring on a finger is just enough of a signal to reduce unwanted advancements.

Rings or other symbols of different sexual states are important for males also. In mating systems where females can have multiple copulations, one male's reproductive outcome can be reduced when the female copulates with other males. From an evolutionary point of view, males in these systems would want to ensure that the female they copulated with is unattractive to other males, or in human terms, the male should put a ring on her finger to dissuade other copulations. So far in this chapter, all of the nature stories have been about females or males becoming more attractive to potential partners. Rings and alternative colors supposedly make females of the species less attractive and, thus, less likely to find additional mating partners or be harassed. Maybe the true knights and damsels of the insect world are the males and females of the green-veined white butterfly (*Pieris napi*).

Both the males and the females of this species can have multiple mating partners. For the female, the benefit of additional partners is increased egg production and increased genetic diversity of its offspring. The male, as stated above, actually has a decreased number of offspring if his mated partner finds other males. Here the work of Dr. Christer Wiklund and his colleagues in Sweden has demonstrated the male's solution to the problem of a wandering female butterfly. Unlike the damselfly mating systems, male green-veined butterflies are quite gentlemanly as copulation can only occur with receptive and willing females. As with most insects, the female emita a sex pheromone that signals to the male butterflies in the area that she is ready to receive visitors. During the mating season, virgin females are courted quite heavily by male butterflies, but only freshly mated females seem to be quickly ignored or abandoned by courting males. During mating in these and other butterflies, the males transfer a sperm packet to the females which is stored for later insemination of her eggs. In addition to the sperm packet, males also transfer their very own special perfume which signals that this female is taken. The perfume is a concoction that acts as an anti-aphrodisiac which serves to repel or dissuade any future male courters. Aphrodisiacs, named after the Greek goddess Aphrodite, are compounds that supposedly increase sexual arousal in the opposite sex. The male green-veined white butterfly has used nature's apothecary to make a chemical that acts in opposition to sex pheromones. Dr. Wiklund tested this by painting the anti-aphrodisiac on to virgin females, and the males treated them as taken ladies. This singular chemical has reversed Hans Christen Anderson's tale and has made the white butterfly (swan) into an ugly duckling.

8.6 Beauty Is Only Nose Deep

Certainly on any given trip across the college campus (or any environment with lots of people), it is possible to see all sorts of mechanisms by which people are attempting to look beautiful, "in style," or handsome. Again, as budding behavioral biologists, I often engage my graduate students in discussions of the latest fashion trends that sweep quickly across a college environment. Close to a decade ago, the latest trend for the fashion conscious student was pajama bottoms and a certain style of boots. As a college professor, I always was intrigued about the idea of rolling out of bed, strolling across campus, and showing up in pajamas for a lecture or lab course. This statement gave way to yoga pants and riding boots. Currently, the modern version of the 1980s sweater dress is slowly creeping back into fashion.

On the male side of things, fashion appears to move much more slowly. Another turn to the 1980s seems to be taking hold of young men as tank tops and mustaches are coming back into vogue. A subtle change noticeable to the discerning eye is the degree of curvature on the bill of a baseball hat. Over a decade ago, curved bills, almost in the shape of a downward pointing C was the correct way to wear a baseball hat. Now, a perfectly straight bill is the preferred way to wear one's baseball hat. A fairly recent phenomenon on the male side of things are scented body washes or body sprays. A series of commercials, clearly aimed at the testosterone driven crowd, makes a series of claims that almost create the concept of a human pheromone.

As we imagine our species with an evolved intellect, one would hope that sex appeal and concepts of beauty travel deeper than superficial aspects like the right pair of shoes or body spray. Among all of the animal stories presented above, the first sign of attraction is through chemical signals. Certainly, animal's final decisions to mate or not are based on suites of signals that employ other sensory channels, but the chemical nature of love and sex appears to be first and foremost among those wanting to mate. As much as we might want to think that we have shed most of our animalistic nature, we are still guided and influenced by the smells around us. Although no known sex pheromone (strictly defined as a chemical evoking mating behavior) has been found for humans, this doesn't mean that our noses don't play a critical role in arousal or attraction.

To understand the effect that olfactory signals have on our sexual attraction, let's explore two different processes that our brains can use to make decisions. For our thought experiment, imagine that we have just entered a coffee shop after a rousing lecture in class (or your favorite food/candy shop after a hard day's work). Since fall is in the air, the special of the day is a wonderfully flavored pumpkin spiced latte. As we enter the shop, the aromas of pumpkin, ginger, and nutmeg as well as the subtly bitter smell of coffee greet us as we walk up to the counter. Having succumbed to the delicious odors, we engage in a short discussion on the size of coffee to purchase. A quick glance at the menu and the prices will allow us a chance to make a decision based on the cost and potential enjoyment we will receive from the pumpkin latte. The decision to purchase a coffee and seek the flavorful enjoyment of pumpkin was purely an emotional decision whereas calculating the cost and the size of our order was a cognitive decision.

Similar distinctions can be made with judgments regarding potential sexual partners. The attraction to a person's physical features or some judgment on the beauty of this person is an emotional decision made using the limbic system of the brain. If asked to judge a person's age, the cognitive centers are activated and a deeper thought process is needed. In a very broad sense, these two processes (emotional and cognitive) also separate our olfactory (and gustatory) systems from our visual and auditory systems. A team of researchers led by Drs. Janina Seubert and Johan Lundström out of the Monell Chemical Senses Center used this separation of processes to ask if human judgment of attractiveness is altered by different odors.

In a familiar experimental design, female volunteers were asked to rate the attractiveness and the age of female models in photographs. This separated the judgment task into a cognitive process (age) and an emotional process (attractiveness). During these judgments, a mixture of odors were delivered to each side of the participant's nose. The odors delivered were either fish oil (unpleasant smell), rose oil (pleasant smell), or some mixture of the two odors. So, the participants were shown an image, smelled an odor, and were asked to provide those judgments about the images presented. The researchers, in an elegant design feature of the experiment, took the photographs and increased or decreased the attractiveness of the appearance with wrinkles and blemishes. This step would help minimize any confounding effect of presenting different models at different ages.

The results of these tests showed that as the amount of rose oil increased in the odor mixture, participants increased their rating on the attractiveness of the photographs. Facial features were more attractive during pleasant odor stimulation. Interestingly, the judgment of age of the models was impacted in two different ways. While being stimulated by pleasant odors, judgments about the age of the models diverged. Younger faces appeared younger and older faces appeared older. Conversely, while smelling the fish oil, this effect disappeared. So, the emotional decision, the one mediated by the limbic system, was heavily influenced by the pleasantness of the odors. There was a positive correlation between the attractiveness of the odors and the attractiveness of the facial features. So far, this effect has only been demonstrated for females viewing female models. It should be interesting to see how far these results extend to males and to people viewing the opposite sex.

This study clearly demonstrates that human emotional decisions and judgments can be guided and influenced by the aromas within our environment. Does one meet a date in a coffee shop or in a loud sweaty bar full of dancing people? The choice of perfume, body spray, cologne, or deodorant may enhance how attractive others think you look. These studies support the idea that odors influence our emotional decisions and while not exactly the sex pheromone that many commercial Web sites claim exist for humans, our perception of odors and the emotional attachment to those odors may guide our choice of a sexual partner. While we would like to think that our highly intelligent species will judge beauty based on features beyond our outward appearance, the reality is that beauty may start right at the tip of our nose.

8.7 The Blood, Sweat, and Tears of Love

In the 1992 movie appropriately titled "Scent of a Woman," Al Pacino gives an award winning performance of a blind ex-army officer who is depressed and seeking one last set of thrills before he commits suicide. Chris O'Donnell plays a poor prep school student who "baby sits" Pacino's character during his last thrill seeking days. The two actors develop characters that compliment nicely off each other as Pacino's world weary Colonel Frank Slade attempts to enlighten O'Donnell's naive Charlie Simms. During a flight to New York City, a flight attendant delivers a couple of drinks and the blind Colonel Slade thanks the attendant by name. Simms, shocked at how the blind military man knew, asks Slade how he guessed her name. Slade explains how the women's accent combined with the aroma of her perfume allowed him to narrow down his choices. He continues on to discuss the scent of a woman and how the hair is the sensuous focal point for that aroma. To Colonel Slade, the alluring aroma of the hair is all that is needed to determine beauty and, in this case, sexual arousal. Are odors strong enough to influence the human sexual state? The answer according to Dr. Noam Sobel of the Weizmann Institute is a resounding yes for both men and women.

Unlike the ants, moths, and mice in the stories above, humans don't produce a specialized signal that has the capability of driving the opposite sex into a frenzy, at least, this chemical has not been found yet. Still, the lack of a specific sex pheromone doesn't mean that humans aren't sensitive to the moods (sexual and otherwise) of the opposite sex. Natural body odors, such as those released during exercise have a complex suite of chemicals that can serve to provide information to discerning noses. In one particular intriguing study, Sobel's team of researchers conducted an experiment to determine if there was any effect of male sweat on the behavior and physiology of heterosexual women. Male sweat contains a steroidal chemical called androstadienone. This compound is also present in saliva and semen of males. In high enough concentrations, it is quite noticeable and not particularly enjoyable, but these researchers wondered what would happen at lower levels of intensity. Instead of asking a group of men to work out, produce a lot of sweat, and collect this odorous liquid, the researchers just ordered the pure steroid and used this substance as a known and controlled stimulant. The subject participants were taken into an odor sealed room and subjected to a number of physiological measurements (like blood pressure and certain hormonal levels) as well as a psychological survey that determined emotional states. During these tests, subjects were provided samples of control smells as well as the sweat steroid androstadienone. The researchers calculated differences between the measurements done before any sniff and after smelling either the control or steroid as well as comparisons between the control and steroid odors. At the psychological level, participants reported heightened sexual arousal and happier moods after smelling the human sweat compound. So, the smell of sweat actually creates a psychological state of attraction similar to the sex pheromones found in so many animals.

Even further though, the aroma of human male sweat changed the physiology of the female participants. After sniffing the androstadienone, female participants had a significant increase in their saliva cortisol levels. Cortisol is a steroid hormone and is often associated with stress. When we feel stress, our bodies often release stores of cortisol in response. The emotional states related to cortisol are quite complex, but the physiological implications are rather straightforward. Cortisol releases energy sources to ready the body for some course of required action. Within the context of Sobel's study, the possible ramification is cortisol is being elevated in preparation for possible sexual activity. The smell of sweat could increase sexual arousal at the psychological level while simultaneously altering the amount of cortisol circulating in the blood stream readying the body for possible physical exertion. Although participants are certainly aware of the positive mood swings and sexual arousal, changes in their internal physiology may be a hidden by-product of smelling the sweat.

So, here is a clear example of male to female chemical communication in regard to sexual relations, but is this chemical communication bidirectional? Do females also send signals that can alter the physiology or psychology of potential male partners? Certainly, the sweat of one sex can influence the blood of the other sex, and Dr. Sobel was interested in other avenues of communication between the sexes.

Humans have numerous sources of potential chemical signals associated with different physiological states and different emotional states. Daily activity patterns, such as exercise, sleep, or stress, can change the types of chemicals being released from our bodies through sweat glands and other pores all across our bodies. Emotional states can also change the composition of different signals that our bodies produce. Emotional tears have a different chemical composition than the normal tears that our body uses to lubricate our eyes. Humans actually produce three types of tears (at least in chemical composition): Basal, reflex, and weeping or crying tears. Each of these tears has a different physiological function, such as immune protect and lubrication (basal tears), washing out foreign objects such as sand or onion odors (reflex tears), and emotional release including both happiness and sadness (weeping tears). This last set of tears appear to be the ideal candidate for potential chemical signals because of the close connection between the production of tears and emotional state.

One of the benefits of working with humans as experimental subjects is that they can communicate their thinking or emotional state. So, in a very simplistic design, researchers can present subjects with a particular odor and essentially ask "How do you feel now"? This descriptive belies the complexity of experimental design and the need for tight and well-thought out controls, but the general sense of this design is correct. Dr. Sobel ran just such an experiment using women's tears as the odor source and asked about the sexual arousal of men. Female participants were shown different video clips or movies that had emotionally sad endings to them. In addition, the participants were asked if they were easy criers, and only those participants who answered yes were asked to participate in the study. In the first of two different aspects of the study, men were shown pictures of women while smelling either women's emotional tears or a saline solution and asked about the emotion that the picture presented and (for a different picture) whether the picture was sexually arousing. Smelling the women's tears did not impact the ability of the men to attach an emotion to a picture, but lowered the feeling of sexual arousal of the other photographs. In a second, more detailed aspect of the study, male participants were asked to lay down in a functional magnetic resonance imager or fMRI. The fMRI allows the researchers to see which parts of the brain are active during different activities. While lying in the fMRI, the men were shown sexually explicit movies and then asked to smell the women's tears. As in the psychological response to photographs, the areas of the brain associated with sexual activity were suppressed by the presence of the tears. Just to check whether the men noticed any difference in the smell of tears or saline solution, the researchers did a series of controls to check for this. The presence of either saline or tears was not discernable to the men, so they were not cognitively aware of what they were smelling. The conclusion of all of these experiments? Women's emotional tears reduce the sexual arousal in men even when the men weren't aware that they were smelling tears. Even more interesting is that the chemicals present in emotional tears have the ability to alter the brain activity in the men without the men knowing that this is happening. Unlike the androstadienone which is noticeable, the tears are ninja-like chemicals sneaking their way into the brain of the unwitting partner and turning down the emotional knobs of sexual arousal.

Whether the source is artificial, like a perfume, or natural, like sweat or tears, the aroma of love certainly alters any suite of emotions associated with physical or sexual attraction. Our level of arousal is heavily influenced by hormones circulating within our blood system, and the most likely mode of action for these signals is through our limbic system or the emotional seat of our brain. From nose to brain to hormone release, the speed of these reactions and the degree of influence can be quite powerful. The Holy Grail for perfumers is that single or group of molecules that is the human sex pheromone that evokes unbridled passion and love, but the search for that grail is as likely to be as successful Monty Python's search in their movie of the same name.

Chapter 9
Smells Like Teen Spirit

Growing Up in a World of Flavors and Fragrances

My parents are very active in their church and have an intense love of music and singing. These two areas are married in their long-term involvement in their church choir. My mother is a soprano and my father sings as a tenor. Their choir is a small handful of dutiful parishioners that lead the congregation through the hymns. I do not know if this is common among church choirs in the Midwest, but this little choir always seemed to be lacking male participants. The altos and sopranos outnumbered the male voices almost two to one during the years that I grew up. Compounded on top of this disparity, the number of men singing bass was always fewer than those that sung tenor. As a result of these mismatches, my older brother was drafted to sing as soon as his voiced changed. His singing voice ended up being a tad lower than my father's voice. Really a baritone in voice, he was close enough to augment the meager amount of basses present.

I am close to 2 years the junior of my brother and since the male voices in the choir were not multiplying rapidly, I knew that my fate was also sealed when puberty caused my voice to change. This was really not as bad as I make it sound here. Looking backward, I appreciate the family togetherness that this brought about. The choir was full of friendly people and there was a good sense of belonging to a real community. Yet, at that time, I did not have the pleasure of an adult brain, but that of a redheaded awkward teenager trying to fit in at school. In my mind at the time, singing in the choir was not really building my social status.

Roughly a year and a half after my brother's voice changed, that magical time of physiological adulthood hit and my body changed as a result of puberty. My voice began to alter and drop as most adolescent's boy voices do. During this change, my voice became a little lower than my father's tenor or brother's baritone voice. So my place with the one other bass was secured and we doubled the power of that choir section. My brother, having been freed from forcing his voice lower, left the basses and joined the tenors. I believe that my voice ended up somewhere in between a baritone and bass which is probably not possible. I found myself able to sing the low baritone notes and high bass notes. In addition to the ambiguous nature of my voice, I did not inherit my mother's, father's, or brother's singing ability; I am not sure that

P.A. Moore, *The Hidden Power of Smell*, DOI 10.1007/978-3-319-15651-4_9

I helped the cause of my fellow bass. I remember many choir practices where I spent my singing time constantly jumping between the upper bass notes and the lower tenor notes. We had a very kind choir director that tended to overlook my musical transgressions, possibly because I was at least singing some of the bass notes (albeit somewhat off key).

Voice wasn't the only thing changing during this time. The kids around me were at all stages of puberty. The males were starting to grow facial hair and voices were cracking during classes. This time period was the late seventies and the cheesy moustache was a popular addition to the faces of young men. A review of the class photos of that period of time would show different amounts of facial hair and yet, every one of us was quite proud of any little bit of hair that would adorn our upper lips. I think we were all unaware of how truly horrible sparse moustaches or beards would look. These were badges of adulthood proudly displayed like a groups of peacocks with their tails advertising their sexual virility.

For the girls, the most obvious changes centered on the blossoming of their chests. Being a male, I can't imagine that a very public measure of your state of puberty is a socially pleasant thing for young girls to experience. Unlike shaving and facial growth, the girls' experience is not controllable and visible for most all to observe especially around young men with raging hormones. The final experience of puberty that attacks both sexes is the dreaded battle with acne. Just like facial hair, secondary sexual characteristics (breasts), and voice changes, the severity of acne would range greatly across individuals.

The timing of puberty also seemed to have a very wide range across this age group. I seemed to be on the leading edge puberty as I was able to grow a full moustache by the age 11. I do seem to remember a larger young boy named George who seemed to have a full beard by the age of 10. The same spectrum appeared with the opposite sex where some girls had "filled out" far earlier than other girls. Puberty, in humans, can start as early as 8 years old and as late as 13 years old, and some of this variation in the age at which puberty starts is due to the genetics of the parents. Outside of parental genetics, environmental factors such as dietary quality and physical activity also influence when puberty starts. Of course, another very important factor for nonhuman mammals is their odor environment.

Many nonhuman mammals live in varying sizes and structures of social groups and, with social groups, comes differing social odors. Unlike the sexual pheromones described in Chap. 8, these odors are the "everyday" variety that are present because these mammals live in close quarters. The common house mouse (*Mus musculus*) is an excellent example of this case. These animals have quite a flexible social behavior structure that is highly dependent upon the environmental conditions. If found in rich habitats (such as human apartment buildings), mouse populations tend to be hierarchical in nature. This means that there are clear social roles with dominant and subordinate animals. In poor habitats, those with not many shelters or food resources, these mice tend to be territorial and both male and female aggression tends to be heightened. The relatedness of groups also changes across different habitats. Within crowded and resource rich human dwellings, social groups can include both related and non-related individuals, whereas in natural settings, social groups tend to include only related individuals. The significance of the relatedness of the social groups plays a role in the way that chemical cues influence sexual development and maturity.

Unlike civilized human environments where we have isolated environments to deal with our urine and feces, social mammals use their reproductive waste to mark territory, identify individuals, and determine the sex and reproductive status of those individuals around them. Mice are no different and use urine cues to mark territories. Urine also plays a different role in the physical development of those younger animals in social groups. Female mice reach puberty faster in the presence of unrelated male urine then in the absence of male urine. The chemical signal, called the major urinary protein complex, is detected in the vomernasal organ. Named the Vandenbergh effect after J.H. Vandenbergh, who discovered the effect in 1975, the presence of the urinary pheromones causes an increase in the gonadotropin releasing hormone in the young female mouse. This hormone is directly responsible for the onset of puberty in mammals. The accelerated puberty enhances the ability of the young female mouse to mate with a close and presumably dominant male.

These signals can accelerate the onset of puberty even in the absence of the male mouse. Its signature urine is all that is necessary to cause this hastening of puberty in the female mice. Along with the Bruce effect (Chap. 8), the functioning of the female mouse reproductive system is sensitive to a number of different odors in their environment. Pregnancies can be terminated, sexual maturity can be accelerated, and estrous cycles can be synchronized all due to the chemical environment that surrounds the female mouse. All of these signals act through a very sensitive vomernasal organ that ultimately sends signals on to the hormonal centers within the mouse's body. Within humans, there is little evidence for fully functional vomernasal organ, so it is unlikely that similar effects will be found within humans. Thus, unfortunately, all those youngsters going through the various stages of puberty at different rates will have to let nature take its course.

9.1 The Smelly School Bully

Seemingly linked to the physiological development that occurs at the awkward ages of 8–13 is the social development. I, not so fondly, remember the formation, destruction, and reformation of different social clichés, groups, and gangs in the socially volatile times of these ages. Brought together by commonalities that range from the purely physical (size or beauty) to the purely social (sports or academic activities), these groups often had social standings that were created through peer pressure or other psychological means. Just as there were ranges in the physiological development of puberty, there were quite a wide variety of social or moral development phases. A common factor that seems apparent across a number of social situations is the classic school bully. Portrayed in various forms in movies and TV shows for both males and females, the male version of the bully tends to be the physically superior individual whereas the female version of the bully tends to be the one controlling different groups of individuals. Although not entirely true to real life, these individuals often resorted to force for the males or peer pressure for the females to extract lunch money, food, or some other payment in exchange for less harassment. This differentiation of modes of actions across males and females that is a product of popular culture is interestingly reflected in nature also.

Nature is not without its bullies especially during this developmental phase of an organism's life. While lunch money is not at play among animals, accesses to critical resources such as food and shelter are nature's equivalent. Many studies with a disparate group of animals have shown the physiological and psychological effects of physical bullying on animals particularly for the runts of litters. If these animals manage to live to sexual maturity, their reproductive output is significantly lower than animals that were not bullied. Olfactory cues can also play a role in the bullying of animals particularly with mammals.

The lesser mouse lemur is an interesting creature with quite an oxymoronic name. The lesser mouse lemur is not a mouse and is not lesser in size. It is the largest member of the mouse lemurs and to add to the curiosity of the name, it is the largest member of the smallest group of primates. Nocturnal in nature, the lesser mouse lemur has rather large eyes to aid in its nighttime lifestyle. Found only on the island of Madagascar, this lemur can be found in multisex groups during day sleeping in the holes in trees. Males and females both have multiple sex partners during reproduction with the females giving birth to typically two offspring after a 60 day gestational period.

The key to this reproductive system is that both males and females have multiple partners. This simple aspect sets up an interesting sexual dynamic in that males would have a reproductive advantage if other males could not be reproductive. By suppressing the sexual development and reproductive potential of surrounding males, the dominant or bully male has an increased chance of siring more of the offspring in the surrounding females. Conversely, females, needing multiple sexual partners, would have an advantage by promoting the sexual development of as many males as possible. These females are acting in a nurturing fashion by enhancing the sexual activity of the poor bullied males.

As with many social mammals, these lesser mouse lemurs have social hierarchies where there is a single dominant male with multiple submissive or subordinate males. These status distinctions are established and reinforced through aggressive interactions. The fights often occur before the breeding season such that the dominance hierarchies are fully established by the time that females become sexually receptive. At this point in time, the males act as school yard bullies and fight to fully determine who is king of the reproductive hill. As the mating season approaches, which is around September and October, overt physical aggression is reduced among the males, but the aroma of dominant males still functions to bully the subordinate males. Males leave urine cues around nests and territories and as the breeding season approaches, scent marking with urine increases. The presence of the urine delays sexual maturity in subordinate males. The male urine pheromones, as yet to be identified, actually decrease the production of testosterone in subordinate males which inhibits their sexual activity.

Testosterone in the males increases the size and mass of the testes which increases the sperm production in sexually active males. Sperm production is important because the females have multiple matings with different males. Multiple matings lead to what is called sperm competition which is a relatively new aspect of sexual selection theory originated by Darwin mentioned in the previous chapter. Sperm competition occurs when females have multiple mating partners, and the fertilization of her eggs is done by sperm from different males. In the race for the most offspring, the evolutionary advantage goes to the male who manages to fertilize the most eggs. If, within the female reproductive organs, sperms from multiple partners are simultaneously present, then there can be competition among the different sperm or sperm packets for reproductive success. This competition can result in a number of different evolutionary solutions. Animals can use behavioral mechanisms such as the active removal of sperm in subsequent mates that occurs in some insects. Many species have mechanical advantages such as sperm precedence where the mating order biases egg fertilization or even biochemical mechanisms like natural spermicide. Thus, when sperm competition exists, having a larger number of robust sperm can increase a male's reproductive success and when dominant male lemurs suppress the production of male sperm in subordinates, they are chemically bullying submissive lemurs in order to reduce the number of potential offspring.

From the female's perspective, this chemical reproductive bullying is counterproductive. If females have multiple sexual partners (up to seven or so), they want each and every mating event to be with robust and hearty males who are up to their highest reproductive potential. Having one mating with a fully functioning male and six matings with less than vigorous males is essentially wasting her time with the sexually inferior males. So, while the dominant males have a selective advantage using their chemicals to suppress any subordinate males around them, the females want to enhance the fertility of all the males in their area. Just as the males use their urine to reproductively bully males, sexually receptive females use urine to increase the fertility of males in the territory. As the female lesser mouse lemurs become sexually receptive, they increase their scent marking using urine as their source of marks. In the sexually mature urine are pheromones that not only attract males to mate but also increase the production of testosterone and spermatogenesis in males that smell her urine. Thus, there exists an evolutionary arms race between the sexes or at least between dominant males and sexually receptive females in the lesser mouse lemurs. The dominant males are producing bullying urine that serves to reduce sperm production in subordinate males while at the same time, females are producing sexually nurturing urine that has the opposite effect on males.

9.2 The Mean Girls Club

For the lesser mouse lemur, the males are the reproductive bullies, but the status of bully is not limited to only one sex. In the 2004 comedy "*Mean Girls*," the bullies of the school is a small clique of girls called "The Plastics." The movie is, unfortunately, a fairly accurate representation minus the comedy of many different high school situations where clichés form and those outside of that cliché are targeted for retribution or mockery. In "*Mean Girls*," those individuals, particularly the other girls, that are deemed unworthy in either beauty or social standing, are ostracized by the plastics. As often with the formation of groups and group dynamics, those inside the group are offered opportunities or treated better than those outside of the group. In-group members receive special benefits like money, access, and information whereas out-group individuals are bereft of any benefits. It pays to be on the inside.

In high school comedy movies as well as nature, the concept of bullying isn't limited to just one sex. Female animals are just as adept at the control of other females as males are controlling the reproductive output of other males. The evolutionary advantage for females is relatively clear. If other female's reproductive output and even sexual receptivity is inhibited then all or most of the offspring are sired by that singular female. This type of situation can be found commonly amongst social insects where a singular queen is the only female capable of reproduction. For social insects like bees, the reproductive situation is different than in social mammals. In social bees, the queen bee and drones (males) are almost genetically identical. Almost because the males only have one set of chromosomes and genes, whereas the female queen has two sets of chromosomes. The set of chromosomes that the males do have is identical to that of one set of the queen's chromosome. The workers are sexually undeveloped females, and their state of continual "prepubescence" is controlled chemically by the presence of the queen and her pheromones. For most social mammals, all of the members of the group, male and female, have two sets of chromosomes and, thus, vary genetically from each other.

Among the primates, marmosets are a group of 22 different new world monkey species that are generally small and live within cooperative groups. Within the primate community, new world monkeys refer to those primates found in Central and South America (whereas old world primates refer to those primates found in Asia, Africa, and the Middle East). These groups of new world monkeys can consist of 15 individuals, but in most species of marmosets only a singular female breeds. As with most primates, the marmosets have a social hierarchy and the dominant females exert considerable control over group dynamics. Just like the plastics in the movie "*Mean Girls*," dominant females often exhibit aggressive behaviors toward other females to either establish or reinforce that dominant–subordinate relationship. The privilege to breed is a benefit of being the dominant female amongst the marmosets. Yet, the restrictive breeding isn't buttressed by the aggressive displays and actions towards subordinates; a pheromone from the dominant marmoset suppresses the ovulation of the other marmosets.

Among mammals including humans, luteinizing hormone (LH) is produced by the pituitary gland and is necessary for ovulation in females. An increase in LH production induces ovulation whereas a decrease in LH production results in a delayed or suppressed ovulation. Without ovulating, females cannot conceive and, thus, are infertile. If juvenile marmosets are exposed to the odors of dominant female marmosets, their LH production is decreased and they do not produce eggs.

Still, the aroma of dominant females does more than act on the physiology of the other marmosets; the subordinate marmosets do not even exhibit adult sexual behavior. This pheromone then stunts both the reproductive physiology of the marmosets as well as the sexual behavior of the subordinates. If the dominant female is removed from the groups, the subordinates begin to produce more LH and exhibit sexual behaviors, at least until another dominant female arises from the ranks of former subordinates. Among the marmosets, the mean girls get all of the guys and it definitely pays (evolutionarily) to be part of that mean group.

9.3 The Unbreakable Bond

During my school days, I was active enough to be a part of different school groups. I played sports through three different seasons (fall, winter, and spring), was part of the music program, and was smart enough to participate in some academic groups. Each of these groups and their leaders had different rituals or events designed to forge a close knit community among its members. One could argue how "effective" these programs were, but being naïve and wanting to be a part of these groups, I blindly followed along just like a sheep.

The fall season for band meant marching practice which includes a week-long marching band camp at some off site location. We stayed 6–8 students per cabin and were grouped by our different instruments. Playing tuba in the band, I stayed with a fairly rambunctious group of boys. Maybe this is my bias, but for me, two groups (drums and tubas) in the marching band stick out as either a tad more theatrical or crazier than other band sections. Certainly, the drum section is unique because they lead the band on and off the field and are the center piece of rhythm for the entire band. This group often seemed a very close group because if one of them is off rhythmically, it is fairly obvious to the audience and band alike. The drum section always seemed to have some special movements, twirls with drum sticks, or shouts that were unique to them. Carrying a rather large tuba or Sousaphone, even in a large band, makes one quite obvious while marching. In addition, the tuba section often had distinctive movements with the instrument and like the drums, being out of sync is an ultimate sin for a marching band. So, while the rest of the band had down time, the drum line and tuba section had their own self-generated practices that resulted in group bonding.

Within athletics and academic groups, similar self-generated events also lead to a group bond that was designed to create a community with a common goal whether that goal was winning a football game, putting on a great show, or winning a debate. These bonds, one could argue, are necessary to focus groups of individuals toward a common vision or mission. They also created a set of in-group and out-group behaviors where those people inside the group (say tuba section) had privileges or received more magnanimous behaviors that those outside the group. I am not trying to equate the differential behavior actions to the mean girl syndrome discussed above, but simply that those members that are part of the "in-group" receive preferential support or attention. Now, some 30 years later, I attempt to create that same in-group bonding with my laboratory. Having potluck dinners, cookouts, game nights, or celebrating each other's accomplishments is designed to create a supportive atmosphere. For anyone going through undergraduate or graduate school knows that this support is almost a requirement. My laboratory group is small enough that there is no need for additional mechanisms for recognizing group members beyond facial features, but we do have numerous lab t-shirts that bear the initials "LSE" for the Laboratory for Sensory Ecology.

As outlined in Chap. 6 in the social dynamics of life, animals have different groups that result in differential support or efforts exhibited toward those inside of the group definition over those outside. The most fundamental distinction for in- and out-group recognition is that of kin. Within kin recognition, the genetic bond connecting groups is permanent as opposed to the participation in a band, team, or graduate lab. The ability to recognize and provide differential support to those that are related and unrelated is central to any concept of social behavior. Narrowing this view even further, the recognition of offspring (and the genetic relationship) can be critical to providing parental care. There are animals that have evolved mechanisms to exploit this differential care, such as the cuckoo bird covered earlier. The cuckoo bird lays its eggs in other birds' nests and without the lack of the genetic relationship between parent and offspring, the other bird raises the cuckoo chick as its own.

Thus, the recognition of one's own offspring and the ability to give support to this offspring and no other can lead to success growth for that offspring. Unlike lab t-shirts or team uniforms, the odors of identity can provide a lifelong connection from mother to infant, and sheep have shown researchers how this connection can be made. As with a number of scientific discoveries, people, and in this case shepherds, were performing this bonding trick for years without even realizing what was happening at the biological level. Shepherds recognized that the bond between mother and lamb is so strong that the mother will not allow any other lamb to suckle from them. This bond happens in the first few hours after birth of the lamb. Thus, any lamb, whose mother died during birth, would also be a lost cause because of this critical bond. Conversely, any dam who has lost her lamb would have an udder of good milk but no lamb to suckle with. To create an artificial bond between foster lamb and mother, the shepherd would take the amniotic fluid from the lamb who passed away and rub the foster lamb in this fluid. Next, the shepherd would present the foster lamb now covered in the other lamb's fluid to the original dam. Once the dam cleaned off the lamb, she would accept this offspring as her own and raise it to adulthood. Shepherds would call this practice rubbing the lamb in "the waters" (amniotic fluid) and "cleanings" (placenta) and would have considerable success using this technique.

Dr. Federic Lévy from France has discovered an intricate series of neurological and behavior events that eventually lead to the dam–lamb bonding. This bond leads from birth canal to the brain and ultimately to the behavior of the mother. During birth of the lamb, the vagina and cervix of the mother sheep gets stretched. The stretching, which can be painful, excites nerve endings and sends these pain messages to the brain of the dam. The brain, upon receiving these messages, triggers the hormonal centers of the body to produce oxytocin which is a hormone that functions during childbirth. The oxytocin finds its way into the olfactory bulb of the mother. The olfactory bulb is the "mini-brain" of the olfactory system and is responsible for the processing and decision-making on olfactory signals. When oxytocin floods the olfactory bulb, the sheep experiences a switch in behavior. Without oxytocin in the olfactory bulb, female sheep are repulsed by the smell of amniotic fluid, whereas with oxytocin causes the mother to be attracted to the smell of amniotic fluid. This is not unlike the strange olfactory and gustatory cravings that human mothers have before childbirth. In addition, the oxytocin increases the dam's ability to learn individual odors. So, the physical act of childbirth sets in motion the ability of the dam to be attracted to and learn her offspring's odor.

Yet, all of the neurological changes are not finished yet. Since the dam is attracted to the smell and taste of the amniotic fluid, she cleans the young lamb and by doing so begins to imprint her recognition of the lamb on those tastes and odors she is receiving. While she is cleaning off her lamb, the stimulation of these chemicals causes neurogenesis in the olfactory bulb. So, these chemicals cause new neurons to grow inside of the olfactory bulb, and the neurons that grow are selectively tuned to those unique odors that are her lamb's chemical signature. Almost the entire olfactory part of her brain is rewired to be attracted to and recognize her lamb by giving birth. The shepherds, unknowingly, took advantage of this entire neurological sequence by rubbing foster lambs in the single most important stimulus to trigger this sequence of events. These odors and the rewiring of the brain forge an unbreakable bond between mother and lamb in sheep and without these chemical connections, the lamb would not receive the parental care necessary to grow to adulthood. Although the exact neurological events in humans are not known in this detail, a similar mechanism is most likely important for that mother–baby bonding that occurs in humans (Chap. 4).

9.4 Waiting for Superman

It is probably not too far of a stretch to say that aromas are the ninjas of the sensory world. As the title of this book seems to indicate, the natural perfumes of the world are hidden, at least consciously, and work their magic on our brains just like ninja's pulling off a secret attack in the middle of the night. The mother sheep example above demonstrates that odors are capable of rewiring the brain and altering the behavior of mother sheep. Even beyond mammals, odors can alter the physiology and neurochemistry of the receivers. In crayfish, just the body odor of a dominant crayfish can turn the crayfish exposed to that odor into a subordinate animal. Dominant and subordinate behaviors in crayfish are controlled by the neurochemical serotonin which is also associated with similar behaviors in humans. Although the exact mechanism that drives this behavioral phenomenon is still unknown, the most likely answer is that the dominant odors influence the production of serotonin in the receiver and by altering brain chemistry, turns a crayfish into a submissive animal. Since the subordinate animal is at a competitive disadvantage to the dominant animal, these odors are altering brain chemistry in the receiver in a detrimental and uncontrollable way.

Crayfish aren't alone in how odors can alter important physiological factors. Maine lobsters, close cousins of the crayfish, are also quite sensitive to environmental odors. Lobsters, like all crustaceans, carry around a hard exoskeleton that serves as armored protection against other lobsters and predators. While this shell works well to protect these animals, the down side for these animals is they need to molt these shells in order to grow and reproduce. As mentioned in Chap. 4, during the process of molting, the animals grow a new shell underneath the older hard shell. This new shell is soft and pliable and doesn't harden until it is exposed to seawater. In order to grow, the lobster creates a slit in the outer hard shell right at the juncture between the main body and the tail. Once the slit is created, the lobster performs an escape maneuver that would make the great Harry Houdini proud. The lobster pulls its entire body, including the soft and smaller new shell right through this slit. Once free from its old body, the lobster draws in water to "inflate" its soft new shell to provide room for growth. The lobster remains soft and vulnerable for the next 30 minutes or so. During this and only during this very short time period, the female lobsters can be impregnated because the males now have access to transfer sperm packets. When not in the reproductive period of the summer months, female lobsters tend to synchronize their molting where all of the animals shed their old exoskeletons in a relatively close period of time. This synchronization is very reminiscent of the McClintock effect in humans where the famous psychologist Martha McClintock found that human females living in close proximity began to synchronize their reproductive cycles.

Lobsters, as with other crustaceans, are a weird combination of social/antisocial animals. When the term social animals is used, images of large charismatic mammals traveling in family groups or herds are often brought to mind. Yet, nature is decidedly not mammalian or vertebrate in reality. At present time, there are roughly 5500 known species of mammals which pale in comparison to the 40,000 known species of crustaceans. At a broader level, there are approximately 60,000 species of vertebrates and over 1.2 million species of invertebrates. These numbers provide some justification for extending definitions of sociality or social behavior to more than just one half of a percent of the animal kingdom. Lobsters exhibit social behavior in that they form dominance hierarchies through ritualized fights. The dominant lobster then exerts and re-exerts its dominance on a daily basis by walking around its territory and temporarily kicking all of the male subordinate animals out of their shelters. Essentially, the big lobster on campus is acting like a bully. Still, it pays to be the Superman of the lobster world because all of the females want to mate with the dominant male. Yet, if all the females are molting at the same time and they can only mate during that precious 30 minutes between molting and hardening of the exoskeleton, there exists a conundrum.

Female lobsters can sense the dominant lobster by its body odors and probably through some specialized chemical signals in the urine. Lobsters love to live in shelters which amount to small dens or holes dug in soft sediments typically located underneath rocks. The females within a territory will roam smelling the water in order to determine which local male is the superman and worthy of her attention. As the summer waters warm and the females begin to move into a molting status, the drive to mate with the dominant lobster increases. So, the females increase their periodic olfactory forays into the male lobster's shelters just to be present when they molt. If the females are ready to molt, they'll actually move into the male's shelter and temporarily live with the dominant lobster. When she molts, the lobsters have sex and she'll stay another 7 days in which the male provides some protection against predation. The conundrum with this system arises if another female wants to mate with the dominant lobster, but he is still cohabitating with his previous partner. As noted above, most of the females molt at the same time of the year.

Molting is a physiological phenomenon controlled by hormone levels in the lobsters, yet it turns out that the timing of the molting is somewhat controllable. If a dominant male is currently tied up with another female, the other females, all ready to molt and mate, will continually come by the Superlobster's fortress of solitude for a good smell of the aromas. If she senses that the mating has ended and the previous female is ready to move out, she begins the molting sequence. If there is no chemical hint that the cohabitation is ending soon, the female lobster has the ability to delay her molting until superman is no longer busy. This has been called "serial monogamy" in that the females have the ability to literally line themselves up in week-long intervals just in order to mate with the dominant lobster. In absence of those dominant male odors, the females all molt at the same time in the summer. When a fertile and dominant male is present with all his aromas, the female lobsters alter their reproductive physiology just to mate with him. Occasionally, if the lineup is too long and one female just can't hold it any longer, she'll rush over to the Jimmy Olsen (Superman's weak sidekick) of the lobster world to mate because mating with a subordinate lobster is better than no mating at all. Waiting for superman is the preferable route for lobsters, but alas, sometimes, superman is just not available (Fig. 9.1).

Fig. 9.1 Lobster deli line waiting for superman

9.5 Fear and Loathing in Odors

The presence of a real superman, not a lobster version of a superman, would provide a sense of security and safety akin to what the females lobsters need during their molting process. In lobsters, the father is essentially gone right after insemination and while the mother provides some parental care and protection after the larva are born, the juvenile lobsters are on their own to find food and avoid predation. While humans aren't under attack from wolffish, cod, or flounders like lobsters, having an excellent sense of what to fear and what not to fear is essential to navigating the often perilous world of the young teen years. Mammalian offspring are altricial offspring which means that they are essentially helpless after birth. Extended parental care is standard for most mammals and the roughly 15–16 years that humans provide care and support is long even among mammals. During these juvenile and teen years, the transfer of knowledge about how the world works can be critical to the survival of the offspring.

During my younger years, my family would often spend our summer vacation camping among the more undeveloped and rustic areas of the midwest and as I got older, we began to explore some of the western states. During these years, I was quite the precocious youngster with little fear of exploring deep woods, streams, or the lakes surrounding our campsite. I am not sure whether my lack of fear of these areas resulted from naiveté or stupidity, but I sure did get into some adventures. I am sure that these adventures are what prompted me to become a biologist as I had this innate curiosity for watching nature and wondering why animals did what they did. Walking through the woods, I would often find holes or burrows in the ground and being curious, I would poke around near the burrow or occasionally stick something down the burrow hole in hopes of coaxing the owner out into the open. Not a particularly bright maneuver for a young budding biologist, but sometimes curiosity gets the better of our logical judgments.

One of these times that I was wandering in the woods looking for burrows, I was accompanied by one of my uncles. This uncle spent a considerable amount of time in the military and was, to the younger me, a rather brawny and gruff individual. A hunter and outdoorsman, there wasn't much that would scare him or even cause him to raise an eyebrow. I did not have a close relationship with him, being 1 of 12 uncles, still I felt quite safe walking through the woods and poking sticks in small and large burrows. The purpose of the journey was not for me to find nature to play with, we were on the hunt for the elusive and delicious morel mushroom to augment our dinner that night. So, while my uncle was examining trees and looking under leaves, I was becoming quite bored and started investigating burrows. I came upon one that was clearly freshly dug and my excitement was heightened. I thought maybe I could discover something interesting inside. I grabbed a living branch covered with leaves off of a young sapling and stuck the branch and leaves into the hole. After a couple of attempts, the branch started to move on its own. In a buzz of emotion, I called out to my uncle that I had found something. He turned and his face was a mixture of intrigue and annoyance. I backed away from the burrow entrance, and this small black creature began to emerge appearing quite put off by antics. As the head of the creature emerged, I saw a tuft of white fur on top of its head and suddenly my uncle yelled, "Run boy!" I was shocked into action, turned on my heels, and followed my uncle for a good 100 meters sprint away from the burrow. As I was running, I wondered what dangerous animals could scare my uncle because certainly the size of this vicious predator wasn't intimidating. As we slowed to a stop, my uncle explained that I had found the burrow of the infamous skunk and his fear wasn't based on any potential injury. His response was based on previous experiences of getting sprayed by a skunk and having to deal with the odorous consequences of provoking it.

Without my uncle's help, I certainly would have learned about skunk and their defensive weaponry, but the transfer of knowledge from his experiences to my mind certainly made my juvenile and now adult years much easier. Although my pet dogs don't seem to have learned lessons about skunks, luckily I have not had a close enough encounter to a skunk to be sprayed. The juvenile years of our lives are full of examples where adults and mentors are preparing us for the good and bad of adulthood. During the younger years of their children, parents are often saying "No" or "don't" as kids begin to explore the world around them. Whether the admonishment is aimed at the danger of streets, unfamiliar dogs or campfires, the goal is to transfer that knowledge of potential danger before a first-hand learning experience happens.

For humans, this knowledge can be transmitted through parental commands or through images such as stop signs or warning notices. Other mammals do not have the benefit of stories, books, or communal knowledge, so learning takes place through other sensory channels. The smell of fear is not a phrase invented for movie titles and funny lines; mammals exude a different type of aroma when placed in situations that evoke fearful responses as opposed to neutral situations. What is interesting about the smell of fear that mammals produce is that juveniles can learn to avoid or fear those same situations just by smelling the fear of their parents. Instead of shouting "Run boy!" with words, other mammals shout this emotion with their pheromones.

At the University of Michigan, Dr. Jacek Debiec led a team of researchers that discovered this pheromonal learning in rats. The researchers constructed an experiment where female rats learned to associate the smell of peppermint with mild unpleasant shocks. After this associated learning occurred, the rat mothers would show a fear response just to the smell of peppermint even without the mild shocks. After this learning, the rat mothers were impregnated and allowed to deliver her pups.

After the pups were born, they were exposed to only peppermint odors under various conditions. The pups were presented peppermint odors in the presence of their mother and without the mother near them. The researchers used a special brain imaging technique that allowed them to monitor neural activity in the brain of the rat pups under all of the various experimental conditions. They used this technique to quantify activity in the amygdala, which is a part of the mammalian brain associated with fear responses. They discovered that rat pups could learn what their mother's feared even without the mother being present during the experiment. Yet, the pups only learned to fear peppermint if the mother's fearful odor was piped into the test chamber.

Fig. 9.2 Fear and ancestry

Peppermint is something that does not normally evoke a fearful response in rats; thus, the rat pups learned to fear something that their mother feared even before she was pregnant and that this learned fear was solely based on the odors given off by the mother. As a follow up to this experiment, Dr. Debiec and his team blocked the activity in the offspring's amygdala and the rats failed to learn this response.

The mother's fearful aroma has the ability to reprogram the pup rat's brain in order to evoke the same emotional response. The transmission of information from mother to offspring about what to fear and what not to fear is done through the pheromones released by the mother. This would be similar to asking my uncle about what a skunk is and learning to fear the skunk based on changes in the smell of my uncle's sweat. While this finding alone is quite alarming, the fact that the rats are fearing an unnatural stimulus (peppermint odor) instead of a more natural odor like a predatory mountain lion adds a level of complexity. Mothers, or at least rat mothers, can learn to fear situations that might be dangerous in their environment before the pups are born and provide that information to their offspring simply by producing the right body odors. The transmission of learned fear and the ability to alter what is feared by learning through chemical signals is approaching the definition of a social language. At the very least, a knowledge of environmental dangers encoded through chemical signals. Whether that danger is wolffish (for lobsters), hawks (for rats), or skunks (for humans), parental fear and loathing is imparted by parents, probably unwittingly through the biochemistry of their bodies (Fig. 9.2).

9.6 Your Ancestor's Taste

Having spent the first 20 some odd years of my life in the Midwest of the United States, I grew up in a household full of Midwestern culture imported from Europe. My parent's ethnic heritage includes some Scottish ancestors via Nova Scotia with a good dash of Welsh thrown in. In addition, family gatherings usually centered on some type of holiday and definitely included large doses of traditional food found in a largely farming family culture. Most meals included some derivation of meat and potatoes and most of that meat didn't deviate far from beef or pork unless deer season was successful or Thanksgiving was at hand.

I am not sure if this is unique to my version of the Midwestern family, but an interesting thing that was missing from most of our meals was a diversity of spices. Salt was definitely used during cooking, and pepper was used sparingly. The use of anything beyond these two basic cooking ingredients seemed to be discouraged, and most certainly any spice that originated outside of Europe was unheard of in my family. As a child and young adult exposed to very little outside of these cooking methods, I was essentially clueless to the large diverse world of tastes and spices.

When my wife and I left home to travel out to Woods Hole for graduate school, little did we know that the trek would take us across the globe in regard to food and spices. During the 4 years between my bachelors and Ph.D. degrees, the lab environment would include a Spaniard from Barcelona, a very stereotypical New Englander from Maine, two students from the Philippines, a German, an Italian, an Argentinian professor, a southerner from Georgia, and a Swiss German. My best friend and mentor turned out to be the German who loved to eat and cook, and my wife and I would spend every Thanksgiving eating a German feast at his small apartment. One summer we lived with the Spaniard and learned the joys of eating potato pancakes at 10:00 pm while dancing the night away. Through this inculcation of different tastes, smells, and foods, my Midwestern taste buds slowly gave way to a palate that included such things as wasabi, cumin, ginger, curry, Weiner schnitzel and, of course, fresh seafood. It was nearly impossible to remain true to the meat and potatoes with no spices diet that I had grown up with and I am quite thankful to that time period that exploded my gustatory world.

Certainly, the parental influence on our taste preferences is well established, and numerous studies have shown that the diversity of food to which we are exposed to early in our childhood establishes the norm for our dietary choices. What if those influences go beyond the normal plate of food placed before our children? What if dietary preferences or taste per se is influenced almost like the smell of fear in the example in the previous section? These are the types of questions answered by Dr. Julie Menella who also resides at the Monell Chemical Senses Center in Philadelphia.

For years, Dr. Menella has been asking questions on how flavors and fragrances influence the youngest members of our society. She is particularly interested in the role that flavors in breast milk can influence taste preferences of babies and later on as those babies mature into children. In an early study, Dr. Menella performed an experiment with two groups of mothers and their breast-feeding children. She asked one group of mothers to consume a garlic pill before breast feeding and in the other group, the mothers consumed a control placebo pill. After this, the research team tracked the amount of time the babies spent attached to the breast and how long they fed. Turns out that the garlic flavor in the breast milk hits maximum concentration in about 1–3 hours after consuming the pill and that the babies could detect the change in flavor of the breast milk. The babies feeding on "flavored" breast milk spent longer feeding than those babies whose mother did not consume the garlic. The second hand flavors in the milk could influence how much the babies preferred the milk and consequently how much nutrition they would receive from each feeding session.

Dr. Menella ran a similar study using carrot juice and found that baby's preferences could be extended beyond just breast milk. In this study, breastfeeding mothers were separated into two groups again. In the first group, the mothers were asked to drink carrot juice while the control group was asked to consume only water. Now, instead of studying the breastfeeding behavior of the babies, the researchers wait until the children were ready for solid food. At this point in the child's life, they were offered two different choices for snacks. One snack was carrot flavored cereal, and the other was just normal cereal. Carrot flavored cereal doesn't sound too appealing, and certainly, the babies whose mothers drank only water, and not carrot juice, thought that the cereal was just a horrible taste. These babies made the appropriate facial expressions showing disgust. Conversely, the babies whose mothers drank the carrot juice showed significantly less distaste in their facial expressions. So, here is another example of how the mother's preferences for food can be transmitted to their children through breast milk.

Still, this just wasn't enough for Dr. Menella, so she extended her study even farther. In an ingenious twist on these experimental designs, she asked pregnant women to drink carrot juice while she asked another group to avoid carrot juice and carrot flavored items. This is called an experiment in prenatal learning. These experiments are the gustatory equivalent of the products that make the claim that playing Mozart through speakers into a pregnant women's womb will stimulate the early brain activity of children and produce the next Einstein or Mozart. The evidence that playing music to children in the womb will produce smarter children is rather suspect, but the opposite is true of the evidence for the transfer of taste preferences during prenatal times. Babies born to mothers who drank carrot juice while they were pregnant showed an increased preference for carrot flavored cereal as opposed to those babies whose mothers avoided carrot food and drink items. Cultural preferences for food is certainly a learned phenomenon. Even now, despite the my current love affair with Cajun food or sushi, I still have craving for that old time steak and potatoes meal of my ancestors that I spent so many years consuming when I was younger. While this craving may have been learned, there is certainly an aspect of my ancestor's palate that was passed down to me during my prenatal time. Just like my red hair and stocky stature harken to my Scottish ancestry, my taste is my ancestor's taste mixed with a dab of my cultural environment and a dollop from the environment of the womb during those critical prenatal months.

9.7 Jonesing for That Smell of Junior

While many different studies, including the one described previously in this chapter, have shown that parents communicate with their offspring with words, actions, and aromas, one wonders if that communication pathway is bidirectional. While parental care is virtually absent among many different species including the lobsters described earlier, parental attachment and affection can be important for the survival of some offspring. For species with altricial offspring, nurturing during the young and adolescent years can determine the likelihood of survival to adulthood. Certainly, nature's parents have a vested interest in keeping their offspring alive and raising them to an age where they are self-sufficient. Despite this interest, the offspring have a bigger interest in staying alive until they get to reproduce. Without sounding too callous, the parents often have other reproductive opportunities as larger bodied animals have multiple mating seasons. Offspring, on the other hand, are a unique combination of genes from their mother and father and have only this one chance to survive to reproduce. Thus, junior definitely wants to ensure that mom and dad stay around to help raise them.

In Chap. 4, I highlighted research that showed how the smell of a newborn baby can excite the pleasure centers of the brain to promote bonding between the parent and child. The bonding ultimately ensures that the child is properly cared for because the parents have an emotional investment in the welfare of their progeny. This explanation certainly makes sense within an evolutionary framework because the offspring carries at least half of the genetic makeup of the parents. From this perspective, the parents certainly benefit genetically and therefore, should want to care for their children as best as possible. What if we flip our thought process on this research and instead of thinking about what benefits the parents received, we focus on the needs and benefits of the child?

Instead of a potentially loving and bonding experience, what if the child could produce an aroma that actually manipulates the parents into taking care of it? Instead of viewing that new baby smell as a source of pleasure for parents, it is possible to view the perfume of a new child as something stronger than pleasure: addiction. Dr. Johannes Frasnelli, Montreal University, has investigated the neural substrates for the olfactory bonding of child and parent. In an all-too-familiar experimental design, the research team presented clothing that has been recently worn by infants to two groups of women: one group of women that recently gave birth (6 months) and another group that had never given birth. The babies, that wore the pajamas provided to the both group of women, were unrelated to either of the two test groups. The researchers were measuring the reaction of the women and any potential changes in brain activity to completely unrelated and unfamiliar infant odors. The caudate nucleus is that part of the brain that is central to pleasure, but is also a critical component of the brain that plays a role in addiction. Addiction can be viewed as a much more intense version of reward learning. For example, as I train my canine companion, Cedric, to perform appropriate behaviors like relieving himself outside or chewing on his toys instead of shoes or clothes, I provide him simple dog treats when he performs positive behaviors. The taste of the treat activates his taste buds which send signals to his brain that eventually reach his caudate nucleus. Once there, the release of certain neurotransmitters signal creates a state of happiness which serves to reinforce those positive behaviors. So, I actively stimulate these neural pathways through the use of dog treats in order to manipulate (or the more pleasant word of train) his behavior to what I want.

If this reward learning systems goes into overdrive, that pleasure seeking system shifts from seeking to craving. The person (or dog) changes their behavior from wanting those treats to needing or desiring those treats in order to continually activate

Fig. 9.3 First smell is free

that pleasure system. The same neural circuits involved in happiness and reward are also involved in craving or addiction. This is the neural circuit involved in seeking alcohol, tobacco, or even our obsession with certain drugs.

Newborn odors sneak their way into these pleasure centers of mothers, but not women who have not had children. These aromas activate the same neural circuits that are out of control for compulsive coffee drinkers, chain smokers, and other addicts. Given that this part of the brain is not turned on in women, who have not had children, probably indicates that the act of childbirth with all of the hormonal changes that accompany the birthing process either alters the pathways within the brain or at least makes certain pathways more susceptible to increased learning. If mothers become "addicted" to the smell of babies, this craving has the potential to ensure that the mother does everything possible to take care of the baby. This olfactory yearning may also explain why mothers with grown child still love to hold and take care of their children's babies or even just love the smell of newborns. They could be jonesing for a jolt of an aroma-mediated happiness that hasn't happened in a while (Fig. 9.3).

9.8 The Odor of Maturity

As described at the beginning of the chapter, many aspects of our bodies change during that magical time of sexual maturity or puberty. Having completed this pathway many years ago, I watched as a spectator as my own son went through many of those same changes that I highlighted previously. The increase in body and facial hair is probably the most visually noticeable, but as virtually any parent of a teenage boy would attest to, the most noticeable change is the olfactory aroma of the bedroom. The unique, and hardly attractive, perfume of the pubescent boy is an interesting mixture of suspect hygiene and body odors being shaped and altered by the hormones associated with maturing.

Most of that unique eau de boy is caused by a toxic mixture of sweat glands fueling a particularly pungent microbial flora in the underarms. Human have sweat glands called eccrine glands. The glands are actually located all over the body, but have a high density in the underarm area. These glands produce a slightly salty solution (that we call sweat) that is used to cool the body as the sweat evaporates. The sweat itself is odorless and is uninteresting when describing the inside of a teenage boy's room. The actual odor is produced by the bacteria that use the sweat as a food source. At the age of puberty, the eccrine glands are joined in the job of sweat production by the apocrine glands. Apocrine glands are located near hair follicles and as hair growth increases during puberty, apocrine glands are activated. In animals, these glands can be associated with pheromone production that was outlined in other chapters. In humans, these glands secrete an oily or waxy substance that is diluted by the sweat produced by the eccrine glands. Again, just like the salty sweat of the eccrine glands, the waxy substance produced by the apocrine glands is not that smelly itself. The real culprits of that teenage boy smell are the bacteria that consume this new type of sweat and as a byproduct of having meals of the apocrine sweat, the bacteria produce that all too familiar stench of the pubescent teenage boy.

Having a teenage boy in my house transported me through an olfactory portal to my teenage years which were spent, in large part, playing some form of sports. Whether the sport was cross country, football, wrestling, baseball, or track, all of these sports required two things that seemed to deal a death blow for fresh air. First, they required large amounts of teenage boys and second, these teenage boys were producing copious quantities of apocrine sweat for all of the bacterial colonies around a locker room. This odiferous production seemed to peak during our wresting season. Our practices were often performed in rooms with elevated temperatures, and wrestling teammates most certainly increased our wrestling skills as well as spreading around all of that sweat. To be honest, we did attempt to keep the mats and towels as clean as possible, but the diligence of high school wrestlers is probably not as good as it should be. The locker room, our gym bags, and uniforms were certainly washed at regular intervals, but there always seems to be some sort of olfactory signal that could guide visitors either toward our practice, or if the visitor was smart, a direct bee line away from the offending source of that pheromone. Although I had a decent set of wrestling skills, intense competition required using all of the weapons in our training arsenal, and one of those weapons was teenage chemical warfare. If I had an exceptionally difficult match in an upcoming tournament, I would conveniently "forget" to wash my uniform for a couple of days and conveniently leave that uniform zipped up in a gym bag in the trunk of my car. After having ripened, I would gladly wear the uniform as it seemed as if I quickly adapted to the presence of the unique odor. At the start of my next match, I would literally lock up with my opponent which means that one of my hands would be wrapped around his neck and

I would draw his head in close to my uniform. Once he inhaled the rather foul aroma, there would be a slight hesitation in my opponents thinking (or even more obvious response that was often demonstrated by my mother as she opened my bag to wash my uniform). I would take advantage of this olfactory distraction to take him down to the mat to the pin. The trick was certainly successful and would lead to a couple of good wins for my career. Now that I am older and perhaps a bit wiser, I don't quite have that teenage odor anymore, although given the age of my son, the smell of teen spirit is certainly alive and well within the house.

Chapter 10
Human Chemical Ecology

The Revealed Power of Our Odorous World

I find myself in the corner of the local coffeehouse. This store is located on the campus of the university, and I come here certain mornings to write and think, at least when I am not at the bakery. In between surges of writing, I like to pause to think about the next paragraph, section, or page that needs to be written. As I mentioned in Chap. 6, I like to watch social interactions among people which is probably driven by the behavioralist that I have become. So, pausing to contemplate thoughts in a coffeehouse lends itself to the observation of dynamic human social groups. This coffeehouse goes through incredible waves of activity and at times, I feel sorry for the baristas who are attempting to keep up with the incredible and periodic influx of customers. The flux in patronage is as predictable as Old Faithful because the driving force is the timing of the class schedule on campus. On most days, the classes at this university start at the half hour and end at 20 minutes past the hour. Starting at 21 minutes after the top of the hour, a slow trickle of students increase into a downpour of coffee drinkers. At peak times, the line is easily 50 people deep, and it is this 50 person social group that often catches my attention.

I view the patrons not as people but as a landscape of tiny bodily factories releasing plumes of person-specific perfumes. Some of these plumes are originating from artificial sources added to our landscapes like perfumes and body sprays, whereas other plumes are quite natural such as those chemicals produced as a result of the genetics and physiology of the body interacting with that person's diet. If I had those chemical spectacles that I imagined back in Chap. 1, I could slip them on and correlate the behavior that I observe with the signals being released by the people around that one person I have chosen to focus on. As the line flows and ebbs toward the counter, the spacing changes between people drawing some odors toward waiting noses. As the people enter the store proper from the main dining area (the coffeehouse is located inside of our student union) they pass under ceiling fans that serve to distribute and mix the tiny plumes among the different people. I can see one particular person who appears not to enjoy the chemical signals being released by the customer in front of him. There is a slight wrinkle on the skin around the nose, and the corners of his mouth turn down to form a frown. As the line moves

P.A. Moore, *The Hidden Power of Smell*, DOI 10.1007/978-3-319-15651-4_10

forward, he attempts to increase the gap between him and the slightly offensive odor in front of him. I can almost observe an intense struggle inside of his brain. The social pressure of reducing that spatial gap by moving the line forward (despite the fact that this will not quicken the other people's service) seems to run counter to the need to create distance between himself and the intriguing olfactory stimuli in front of him. This singular interaction is played out against the aroma of roasted coffee, flavor additives, and pastries. Despite the couple of hundred little plumes of odors swirling around this single individual, he is able to identify the source of the offending stimuli and attempt to move himself away from this source.

The scientific study of these behavioral interactions coupled with an understanding of the chemical stimuli that influence those interactions is called chemical ecology. Chemical ecology, as a scientific discipline, is rather broad and includes scientists that just study the production and structure of chemical signals, scientists that understand the mechanisms of how signals are received and encoded by the nervous systems under study, and scientists that study the behavioral and ecological interactions that are regulated or influenced by chemical signals. Given this large spread of scientific expertise, we are a loose collection of individuals with very different expertise and interests. At one end of the field, there are chemists working hard to identify, purify, and create the thousands of chemical signals produced by the microbes, plants, and animals in nature. In the middle of this imaginary spectrum are physiologists and neuroscientists attempting to understand how these chemicals are perceived and encoded by the sensory systems and how these chemicals can alter or enhance physiological pathways within an organism's body. Finally, at the far end of the spectrum are the behavioralist, ecologists, and evolutionary biologists placing these interactions among organisms and their chemical world in some broader context.

While the field of chemical ecology is a relatively young field compared to other biological disciplines, the field of human chemical ecology is virtually nonexistent. Most certainly, there are very capable researchers (some covered in this book) working on human olfaction and taste as well as human behavior, but the field as a whole is missing an evolutionary and ecological perspective that is present when we study nonhuman animals. Part of this missing piece is historical in that human sensory research has been focused on vision and hearing. These are our dominant senses and medical research to repair the functioning of these senses has been a major focus of funding and effort. Part is due to the difficulty of designing nicely controlled experiments using humans. For obvious ethical reasons, I can perform experiments on my crayfish that are just not possible in humans. Something as simple as isolating crayfish to remove any social history is ethically wrong for any study on humans. Finally, the field of chemical ecology has lagged visual and auditory ecology because the stimulus is exceedingly difficult to control and measure. As detailed in Chap. 2, because turbulent wind and water flow are essential for delivering chemical stimuli, the replication of stimuli from trial to trial is physically impossible. Despite these limitations on research and theory, it is quite possible to draw together the field's findings on humans and other organisms to more fully develop a field called human chemical ecology.

Human chemical ecologists would be charged with understanding the role that chemical signals play in guiding our behavior as we move through our habitats. Although it may be possible to think of our ecology as somewhat artificial as we read books inside of environmentally controlled buildings and check our email over small handheld devices, this is the ecology that we have built for ourselves. Our environmentally controlled dwellings are fundamentally no different than those air conditioned termite mounts or underground burrows that remain fairly constant in regard to temperature. Our social interactions, as briefly covered in Chap. 6, are fundamentally no different than those interactions among the various primates or mammals found throughout the global. For all of these ecological interactions, we require information that is acquired through our sensory capabilities, and our noses and tongues should not be ignored as powerful sources of information. How we smell and taste our way through our lives should be revealed in the form of a broader context of human chemical ecology. The complex and wondrous manner in which odors guide our lives is truly breathtaking.

10.1 The Human Smell

In his 1992 scientific treatise, D. Michael Stoddart called ourselves "The Scented Ape" which is quite an appropriate moniker for our species. Despite the presence of skin which functions to ensure that certain molecules stay inside of our bodies and others are not allowed to enter, we are a porous bag of chemicals. Our bodies are covered with numerous spots where our internal chemical composition is released into the world around us. The most obvious spots are ones that we tend to cover with perfumes and colognes such as the armpits, the nipples, and chest. In addition, the genital area of both females and males produces a suite of odors that often reflects our internal physiology including our reproductive status. Finally, a number of less obvious areas are responsible for the entire umwelt that is our personal odor. These locations include the skin around the nose itself, our scalp, our hair, the belly button, forehead, ear canal, and even feet can be sources of distinctive smells (Fig. 10.1).

The organ system most responsible for the production of our odors is called the endocrine system. The endocrine system is a collection of glands and organs that periodically release hormones into our blood stream. Contained within the endocrine system are sebaceous and axillary glands, salivary, and adrenal glands as well as many of those organs associated with sexuality including the ovaries and testes. Both of the major skin glands (sebaceous and axillary) produce a diversity of chemicals that read like some weird chemical ingredients in some highly processed food. Mono-, Di-, and Triglycerides, waxy esters, sterols, and fatty acids can be found within these glands and are routinely secreted onto our skin.

What exact chemicals are released and when they are released are dependent upon the ecological and social situation we find ourselves, our age, and our sex. Men and women have different distributions of these glands across our bodies and it appears as if women have a significantly higher number of sebaceous glands on

Fig. 10.1 Da Vinci man

their bodies compared to men. What seems like an obvious statement to anyone associated with kids as they hit puberty, the chemical output of these glands increases significantly as humans hit puberty. This might explain that peculiar aroma that is often found in the vicinity of teenage boys rooms. The production of these glands tends to peak as humans hit the twenties. As our body ages even more, the chemical production of these bodily factories decreases and alters slightly which may explain some of the aromas associated with the elderly.

Social situations and our emotional reaction to those situations can alter how our bodies produce different odors. If we are faced with one of our greatest fears, public speaking, the subsequent stress response releases a number of hormones causing a physiological change. Cortisol, the ubiquitous hormone for stress, flows through our blood causing us to sweat and potentially (depending on how fearful or stressed we are) triggering further hormonal cascades including adrenaline which prepares us to run away if necessary. These hormonal cascades alter the body's odor which is called the smell of fear. Other emotions including happiness, joy, sadness, grief, all trigger different hormones to be released from our endocrine system which has the potential to alter the body's aroma.

Other aspects of our daily lives can also contribute to the types of natural perfumes we produce. Apart from the noticeable changes in our breath with the consumption of garlic and the particular odor of our urine after consuming asparagus, spices, herbs, fish, and different vegetables can contribute to changes in our physiology that subsequently can be perceived as body odor. A binge night on the town consuming alcohol causes an increase in the liver's production of acetic acid which can be released through our skin's pores. The consumption of sulfur containing vegetables, like broccoli, cabbage, and cauliflower, increases the release of more noxious body odors. The importance of our diet in altering body odor is demonstrated by the many doctors that suggest a significant change in diet if pervasive body odor is a problem.

Our personal smell, that identifying perfume that is as unique as one's signature, is a combination of sex, age, diet, culture, health, and the emotional milieu of a person's current condition. A good portion of these personal perfumes arise from the diversity of bacteria living on our bodies and consuming those glandular secretions. Thus, the real personal odor is the combination of what we produce ourselves mixed with those odors given off by our unique culture of bacteria living on our skin. Given the interconnectedness between all of these factors, it is not surprising that our pets can certainly identify us and our emotional state at any moment in time. Most likely, if we are properly trained, we would be able to surely identify each other by our sense of smell as easily as we use facial features to perform the same task. In a previous chapter (Chap. 5), I wrote about my colleague's father who was a rural physician. He often claimed that the smell of the house would provide as much insight into the health of the patient as any visual diagnostic would provide.

10.2 I Smell Therefore, I Am......

As a human chemical ecologist with an understanding on how our bodies produce and respond to odors, I think it is definitely possible to become the Sherlock Holmes of the odor world. The fictional detective, famous for his visual acuity connected to his impeccable logic, could train his nose as easily as he trained his eye to notice the subtle, yet important, connection between odors and information. Alluded to in Sir Arthur Canon Doyle's writings on Sherlock Holmes is the detective's ability to know at least 75 different perfumes. Also, the detective is famous for his appreciation of the olfactory ability of canines in detecting the perfumes of people and nature. Without the astute mind of Holmes, the question becomes how do these odors inform and influence our daily interactions and more importantly our judgments on people. In a 2006 study, researchers demonstrated that women prefer the body odors from male subjects who had not eaten red meat for 2 weeks. The male subjects were divided in to two groups: one with a diet of normal red meat consumption and one with no meat consumption. To control for potential individual differences in body odors, the researchers repeated the trials at a later date with the same male participants; only their diets were switched. In other words, the meat eaters became vegetarians and vice versa. The results were fairly clear and straight forward. The body odor of the men on the no red meat diet smelled cleaner and more refreshing than the body odor of the other men.

These female participants are making fairly powerful judgments of the person on the other end of the molecule. There is growing scientific evidence that attractiveness of strangers is highly dependent upon the perception of both body and environmental odors. Whether individuals are asked to judge photographs, videos, or imagine the person that would produce the smell being sampled, all results point to a correlation between pleasantness of the odor with ratings of attractiveness. Whether the participant's noses are being bathed in an artificial odor (say rose oils) or cotton containing axillary odors, the increase in sexual attraction is almost universal. Although not a shocking discovery given the amount of money spent covering up or enhancing our personal odors, the most unusual part of these studies is that the participants don't need to be consciously aware of the odors. Asked if they remember or can identify the presence of any odors, most participants indicate that they are unaware of any change or manipulation to their smelly world. Thus, these odors are working subconsciously to alter conscious decisions and cognitively complex judgments.

As discussed in Chap. 4 on personal identification, facial features are the primary mechanism by which we identify each other. The style of the hair, the presence or absence of facial hair, the shape and size of the nose and eyes are all excellent indicators for identifying individuals. Some of these features go through periodic changes as the changing of hair color or style, while other features, such as eye color, shape, or size remain consistent throughout our lives. The constancy of these features makes them excellent sources of information for identifying individuals. Remembering our friends and family would be impossible if our features changed on an hourly or daily basis. Since body odors are in constant states of flux depending on our emotional state and dietary choices, the complex mixtures of chemicals being released by our glands and being altered by the bacteria on our bodies are probably not highly reliable indicators of individual identity.

Instead, these odors are definitely reliable indicators of our internal physiological state. This internal state that alters how the body produces chemicals includes the health of the individual, age, sex, emotional state, or even sexual orientation. Work out of the Monell Chemical Senses Center has shown that preferences for body odors differ based on the sexual orientation of the donor and person making the judgments. The perceived level of pleasantness of body odors for heterosexual males were no different when the donors were either heterosexual males, lesbians, or heterosexual females, but odors donated from gay males were perceived as the most unpleasant. When gay males were asked to judge the pleasantness of body odors, heterosexual females were judged the most pleasant followed by the odors of gay males. Odors donated by heterosexual males and lesbians were the least pleasant. Similar findings were found when the participants were heterosexual females or lesbians. What is interesting about these findings is it is possible to identify certain subcategories of individuals based on body odor alone and that the level of pleasantness is highly dependent upon the sexual orientation of the donor and the receiver. Clearly these findings indicate that odors are playing a role in identifying individuals beyond just name or familiarity.

Emotional states can also be identified through odors. In a couple of chapters, I covered the concept of the smell of fear. Several animals including humans have the ability to detect levels of fear purely through the sense of smell. In other sections of the book, I demonstrated that emotional tears are chemically different than tears that function to lubricate our eyes and that these tears have the ability to alter sexual arousal in males who smell the tears. These are examples of the emotional state of the individual altering the body's physiology which in turn alters the type and quality of chemical signals given off by the body. The internal emotional state clearly produces identifiable signals in the form of chemical cues. To date, research has focused on two primary emotional states, fear and sadness, and the research is clear that these two emotions can be identified by the unique chemicals produced by the individual feeling those emotions. These emotions, and in particular fear, are powerful emotions and are likely tied to an evolutionary history where fear was connected to survival. Factors in the environment that evoked fear were likely associated with predation or dangerous environmental conditions such as storms. Within animal social systems, alarm or warning calls spread through the local population rather quickly and efficiently. So, given the critical nature of the detection and identification of the information in fearful signals, it would not be surprising that the ability to respond and empathize with these signals still exist within us.

Beyond the sexual orientation and the emotional state of individuals, an ability to recognize the state of health of individuals could be a critical piece of information necessary for survival. Sociality and the close contact that sociality brings enhances the spread of disease through populations. Advancements in our understanding of disease transmission have brought about changes that are designed to limit or curtail the spread of germs. Covering our mouth when we sneeze, coughing into our arms, and the use of sanitizers are all examples of behaviors that keep the healthy from being infected. The quick identification of diagnosis of sick individuals could also help alleviate illnesses and their spread throughout the population. As with emotional states, illness alters the body's physiology which creates the internal chemical changes that could create detectable differences in body odors. Research has clearly demonstrated that canines have the ability to detect cancers present within a body off of olfactory cues. Often the canines can detect the early onset of cancer and far earlier than most of the technological capabilities at this point in time. This ability, and potentially our own, is connected to how the body responds to disease.

In response to any illness, our body's immune system kicks into gear and begins to generate antibodies to fight off the disease. The immune system is connected to the MHC complex which was discussed in Chap. 4. The MHC molecules are involved in the interactions between white blood cells and other cells in our bodies. As the immune system begins to respond to illnesses, the MHC system of molecules is also activated. If we have the potential to identify individuals by the MHC system, then the same mechanism involved in individual recognition could be involved in health recognition. A team of researchers at the Karolinska Institute in Sweden have shown that humans have the ability to recognize when individuals are sick based on body odors. These researchers gave participants either an injection of saline or a small dose of a toxin that is not harmful, but does activate the immune system. As with most human studies, t-shirts were worn and other participants were asked to judge the pleasantness of those t-shirts. Body odors produced by individuals whose immune systems were activated were judged more unpleasant than those individuals who were given saline injections. Even more surprising results arose when the researchers did a chemical analysis of those t-shirts. It appears as if the concentration of chemicals does not change with the state of illness (or immune system activation), but that different chemicals are being produced by the body. The chemical signature of a healthy body is different than the signature of a body with illness, and that change in signature is detectable through the production of body odor.

All of these studies clearly demonstrate a singular finding for human chemical ecology. Information contained within the chemical composition of our body odors can be detected and used to make cognitively complex decisions and judgments about those individuals. These odors, uncontrollable by any means, are windows to our internal states and signal direct information to the fundamental nature of the person. In a twist of Descartes's statement of "I think, therefore I am," our bodies are sending signals that state "I smell, therefore I am…sick, sad, fearful, or heterosexual" and we alter our behavior based on that information.

10.3 Making Mister Right

The behavioral descriptions above demonstrate an interesting difference between the fields of animal behavior and behavioral ecology. In essence, the field of animal behavior is more mechanistic and cares about the description of the behavior with an eye to how that behavior arises. The underlying neural circuits and muscles and hormones involved are the important pieces. For behavioral ecology, the ultimate survival impact of the behavior and how the behavior functions within the context of the natural environment is the important point. The fields obviously overlap a good deal, but the word ecology added to the behavior means that researchers are slightly more interested in the evolution of that behavior. B.F. Skinner, a famous psychologist and animal behavioralist, was far more interested in understanding how pigeons learned. So he placed them in a box and asked them to peck away at buttons for food. At the other end of the spectrum would be Dian Fossey who went to the field to study her primates and their behavior in the context of their natural habitat.

So, a field of human chemical ecology would certainly entail the studies above and many of the studies covered in previous chapters of this book. If you have read this far then mentioning t-shirt studies or studies that have placed cotton patches in underarms have become a familiar part of the human section of each chapter. These experiments are all well designed and elegant in their own ways, but aren't very ecological. Although one could make an argument that our cities, towns, and suburbs are quite artificial compared to our very recent habitats in evolutionary history, these habitats are more natural than the laboratory setting of sniffing glass vials or cotton pads. Placing the behavioral responses within the context of our behavior in more natural settings or more natural conditions would be more appropriate of a field of human chemical ecology. For a sexually reproducing species, one of the most important ecological and evolutionary contexts is searching for just that right partner. Animals, plants, and other organisms have evolved pheromones for finding just the right partner, but what if those same odors could make the right partner.

Maybe the most important odor to produce is one that would evoke love and companionship. Certainly the host of perfumes that flood the commercial market and even those t-shirt match maker parties that appeared earlier in the book would speak about the importance of finding just that right someone to spend our lives with. While I am doubtful that there exists a pheromone that acts in a similar fashion as what is found within nature, it is certainly possible to demonstrate several types of aromas that alter the behavior of the person exposed to that odor. As covered in previous chapters, the smell of babies produces interesting results in modifying or evoking certain behaviors and emotions in women. Also, the type of emotion or behavior that this odor causes is dependent upon whether the women have children of their own. This scientific observation of what probably is fairly common knowledge is demonstrated clearly, there exists another set of odors that is involved in at least the conception of and maybe in the raising of children. For a species that reproduces sexually, focusing research on how odors and chemicals can alter or enhance the behavior of mothers is really only half the story. The other half of the story revolves around not only finding, but making Mr. Right.

A controversial topic surrounding male partners and pregnancy is something called the Couvade Syndrome (controversial because the syndrome is not fully established yet). The Couvade syndrome occurs when a male begins to mimic the physiological and psychological effects of pregnancy in his partner. Reports of morning sickness, weight gain, and maternity pains are all symptoms of this syndrome. Research to date has shown that some males have increased prolactin and decreased testosterone during a pregnancy by their partner. Testosterone, as a hormone, does decrease nurturing behavior in both males and females, whereas prolactin, best known for stimulating the production of milk in lactating females, has been implicated in an increase in bonding and nurturing of infants. Though research on Couvade Syndrome is sparse, reports of this type have been increasing. In the book *The Female Brain*, Dr. Louann Brizendine suggests that pheromones given off by the pregnant women may be responsible for the hormonal changes seen in male partners. The connection between pheromones and hormones is certainly universal in mammals and other animals, so a connection between the many potential body odors being produced by a pregnant female and the hormonal fluctuations in her partner is a reasonable hypothesis at this time. If true, then these pheromones are not only promoting bonding and connections during pregnancy, they can actually alter the levels of nurturing behaviors in males making them a more ideal partner.

The ecological significance of this potential finding can be found in the ability to raise our altricial offspring. Altricial offsprings are those that are essentially helpless at birth and need considerable parental investment to survive. Parental bonding and nurturing are essential in humans, and in modern culture, significant parental involvement in the rearing of offspring is performed up until the age of the mid teen years. Raising children with two parents is an easier task than doing the job as a single parent. These chemical signals could be selected by evolution to help bond males to mothers and to the subsequent offspring. The selection of the right mate even if imprecise could be improved by ensuring that the nurturing and caring emotions of the father are increased as a result of the right pheromones.

10.4 Wearing Our Scents on Our Sleeves

As much as we want to pride ourselves on our intellect, we are still an emotional species. Throughout this book, I have described social settings and have attempted to explain some of what my eyes and ears have gathered from being in the presence of those gatherings. Even as I write that previous sentence, I want to write "what I have observed from those settings," but focusing on only a single channel of information (vision) sells ourselves and our sensory capabilities short. Whether walking into work, sitting in a bakery, or being among my students, there are definitely signs that reveal the emotional state of those people around me. Micro-expressions (according to Paul Ekman, see Chap. 7) are those subtle and consciously uncontrollable facial movements that convey love, anger, and disgust. The position of eye brows, arms, legs, and even larger scale posture can provide clues to the state of being of that individual. Auditory signs can include which words are being emphasized, stressed, or glossed over. The volume and intensity (coupled with hand gestures) are clear signals to the level of anger or care that the speaker is currently experiencing.

The laughter of the morning ladies in the bakery coupled with them throwing their heads back sprinkled with the occasional touching of the forearm of their friends creates a scene of deep friendships and a relaxed atmosphere of humor. An intense meeting with power suits and ties, coupled with furrowed brows, and terse words seem to indicate that a level of threat, fear, or anxiety is present among the participants. As people come forward to engage my dog, Cedric, they extend a hand, crouch down (in part because he is only 2 feet tall and in part to make their physical appearance smaller), and have a cheery sing-song voice all designed to impart an air of friendliness. If I had the innate skills of my four footed friend or maybe even the proper training, I could add my nose to the list of sensory apparatus which I use to extract relevant information. I could image having a miniature Cedric up my sleeve to carry to different social settings, so that I had access to his keen sense of smell in order to determine the emotional mood of my fellow humans. We do wear our emotions on our sleeves, and these emotions are readily apparent to most of the world.

As touched on several times in this book, the emotions we feel and exhibit to those around us are closely tied to the chemical milieu in our brain and coursing through our veins. When we are happy, our brain releases a certain suite of chemicals including the neurotransmitters dopamine, serotonin, and some endorphins that trigger the right reward circuits. When we are sad or angry, a different batch of neurotransmitters activate different areas of our brain. Yet, it isn't only our brain that changes as a result of happiness or sadness, the entire chemical factory that is our metabolism alters both the types of chemicals being produced as well as the quantity of those chemicals. Our endocrine system as well as our digestive and immune systems have different chemicals present as our emotions fluctuate. In the excellent book, "The Molecules of Emotion," neuroscientist and author Candace Pert details the myriad changes in the chemistry of the body as we feel different emotions. As our body chemistry changes, our personal body odor also changes. Thus, instead of wearing our hearts on our sleeves, we wear our aromas on our sleeves.

Producing unique body odors during different emotional situations sets the stage for the possibility of determining other's emotional state. Two questions need to be answered before we conclude that our noses are just as reliable as our eyes and ears. First, are our olfactory abilities sensitive enough to detect these differences? Second, do we use this information to make social decisions (in other words, judgments) about those individuals? In the field of chemical ecology, the approach is designed to place both the behavior and the information conveyed in the chemical signals in some type of ecological significance. Demonstrating that an animal is capable of performing an action, say teaching my dog, Cedric, to retrieve a ball, provides some insight into the capabilities of the brain under study. Definitely a worthy scientific study, but if that same behavior, fetching a ball, has little to no ecological significance, then the behavior is an interesting phenomenon and nothing more. Demonstrating that humans have the ability to detect emotions via chemical signals might be an important demonstration of what we are capable of doing. Yet, if this detection of emotions occurs only within a laboratory setting and doesn't result in a change in behavior or a utility for that behavior in broader settings, then little can be drawn from a human chemical ecology perspective. The social relevance of this work has to be demonstrated and the correspondence or communication between individuals needs to be a part of the theoretical thinking on these subjects. Thankfully, some of that work has begun to emerge.

In previous sections of this book, the research on the connection between emotions and chemical stimuli has been explained. Clearly, the smell of babies alters the thought processes and subsequent behaviors of mothers and fathers, lover's sexual arousal is either enhanced or abated by the presence of body sweat or tears, and the smell or taste of mint can produce heightened focus for studying or reading. These studies have laid the foundation for the idea that human behavior is influenced and in some instances controlled by the presence of our surrounding odor landscape. In each of these scenarios, the influence of the chemical signal is targeted toward a specific individual in a one-on-one relationship; the lover, or the child. Although not a part of the design of these studies, the larger social context and relevance of chemical signals are missing from this work. Social communication, that is communication and the behavior those signals evoke, is the next step to demonstrate that our chemical world is far more important in our daily interactions then we recognize.

The difference between sending signals of emotion, love, fear, or anger and actually communicating information used for social judgments is subtle but important for understanding the full breadth of human chemical ecology. Perhaps a thought experiment would distinguish these two scenarios. As in previous work, we could take a group of people and work with them as odor donors. In typical fashion, they would don t-shirts or place cotton pads underneath their armpits and would be given some experience. This experience could be working out, watching a funny or scary movie, or judging the emotional context of photographs. These t-shirts or pads are given to the human test subjects and after a sniff or two, questions or tasks can be performed. This general description of an experimental design covers just about all of the human studies found within this book. If the researchers found that after smelling a t-shirt, the test subject's behavior was changed, then we could conclude that humans are capable of determining the emotional states of the senders. The subtle difference between this general human study and all of the animal studies covered in previous chapters is that there is no demonstration of an evolutionary advantage for the sender. Communication, as defined by science, requires that both the sender and receiver of the signal benefit from that information. Simply picking up on signals wafting through the air can be important for the receiver, but carries no inherent benefit for the sender. Sensing the random chemicals in the air is considered a form of obtaining cues rather than true communication.

If a subtle tweak in the experiment above is performed, we could draw a broader conclusion about communication as opposed to picking up cues from the environment. Let's repeat the experiment from above, but instead of having the odor donors watch a random movie clip that evokes an emotion, we need the donors to actively emote a desired response. Imagine that instead of random donors off the street, we hire trained actors to perform different scenes from plays that would require them to emote sadness, anger, happiness, or some other emotion. After this, we would test the actor's smell to determine if the smeller could determine the emotion being performed. If so, this would demonstrate the communication of emotions through chemical signals. These findings would start an entire field akin to linguistics, but with chemicals and molecules taking the place of written and spoken words. We wear our hearts, not really on our sleeves, but on the molecules being produced when we love, cry, or are afraid. As we release these signals, are others using that information to make social judgments about our abilities and our personalities?

10.5 The Odorous Cloud of Human Judgment

Despite the idea that judgment often has a negative connotation, we, as a species, are awash in a world of judgment. In my own world of higher education, that judgment often involves measuring the knowledge of students and attaching some point total or grade to the demonstration of that knowledge. In addition, we sit on committees to appraise the readiness of graduate students to graduate and tenure for junior colleagues. Outside the academic world, every job interview involves an assessment of the applicant's abilities. These determinations of worthiness range from influential career decisions to something as small as whether to trust a salesman. I would hope that each and every one of these decisions is based on realistic measurements of the person's ability rather than how they look or dress, but I am not naïve enough to believe that. Without too much effort searching on the internet or through the print media, it is possible to find copious quantities of examples where job decisions, as well as other decisions, are based on something other than competence. One of the very first things that lawyers tell their clients if they are approaching a jury trial is to make sure that they look their very best when in the court room. The state of dress should not be connected to the guilt or innocence of a crime, but humans are notoriously influenced by the visual signals of a good outfit or the pleasantness of a voice.

Dr. Pam Dalton, another Monell Chemical Senses scientist, is interested in, among other things, the role that odors play in the social communication between humans. Human social communication is dominated by visual, auditory, and mechanical signals. The firmness of a handshake or the touch of a hand on a forearm all convey certain social ideas. A smile and frown are visual signals designed to give some insight into the emotion of the sender, and the tone and sharpness of our words also convey information on whether we are angry or happy. What is clearly missing from this list, and something that Dr. Dalton wants to correct, is the nose that is plainly displayed on our face.

Dr. Dalton and her colleagues started with a very similar and oft repeated experimental design explained in the previous section. A set of individuals donated axillary odors under stress, non-stress, and exercise conditions and then other individuals were asked to smell these odors. While in the presence of these odors, the test subjects were asked to make the classic judgment of the emotional state of various sets of pictures. All of this design was a necessary control to be able to add an additional twist to the experiment. Not only were participants asked to assess the emotional state of a set of standard facial pictures, they were also asked to judge their level of competence, confidence, and trustworthiness. Previous research, some covered in earlier sections of this book, have shown humans can detect stress and other emotions through sweat. Yet, these three assessments do not have any direct connection to an emotional state of a donor subject. As in other studies, the participants were not consciously aware if which odor (sweat or control) they were smelling. The results of this study show three intriguing results. First, that men's judgments of a person's confidence, competence, and trustworthiness are heavily, and negatively, influenced by the presence of stress odors. Second, that women's judgment of these same three traits were not influenced by the presence of sweat, and finally, commercial deodorants can actually block the negative influence of these odors on men's evaluation.

Probably the most compelling aspect of these findings is the attachment of a skill or competency level as a result of the smell of stress. In an old 1987 commercial for an antiperspirant, the fashion designer Donna Karan states that one of the rules of fashion was "Never let them see you sweat" which now should be changed to never let them smell you sweat. In Chap. 2, I designed in moderately simplistic terms how olfactory information is processed by the emotional centers of the brain before being passed on to the more cognitive centers. With these neural connections, it is not all that surprising that humans can sense emotions through our noses. After all, we emotionally feel these odors before we think about them. Thus, a concept of odor empathy is not too difficult to understand. Making measurements of a person's competency requires that this olfactory information continue on through the emotional centers to the cognitive centers responsible for social judgments: the frontal lobe. The frontal lobe is responsible for decision-making and planning for the future. So it is possible that the molecules responsible for stress odor send tendrils of influence all of the way through our noses (at least the male noses) to the frontal lobe to change, alter, or control the decisions about the person who is sweating. So, maybe that old commercial was more omniscient than the writing knew. Although it is certainly unfashionable to let them see you sweat, it is far more costly in terms of social capital to let them smell you sweat.

10.6 The Cologne of Confidence

In Chap. 1, I mentioned the Patrick Süskind book "Perfume" where the protagonist has the ability to concoct perfumes of such perfection that the aromas can control both the emotional state and the cognitive process of people around him. Dr. Dalton and her colleagues have shown that the cognitive substrates that would allow these perfumes to work exist within our brains. Psychology has demonstrated that certain color combinations are associated with different emotions or with a certain social status. Red emotes power and control, whereas yellow is designed to be uplifting in nature. The same can be said for body language, verbal tones, and all sorts of other stimuli. Standing with the feet about shoulder width apart, the arms crossed in front of the chest, and a stony facial expression sends one message whereas standing with the hands in pants pockets, head slight cocked to one side, and a hearty smile is an entirely different meaning. So, associating certain odors with differential behavioral states is not out of the question. For a number of these visual and auditory stimuli, the meaning of the colors, body language, or verbal tones is most certainly culturally specific and learned as we develop and grow within that culture. The same learning and culturally specific associations could be true for these odors.

On the flipside of these ideas is the ability to hide or mask the body odors that could negatively impact our lives. Highly stressful situations, such as a job interview or meeting the parents of one's significant other, stimulates the production of unwanted and harmful (from a social status perspective) odors. The moment that you reach out to shake the hand of your girlfriend's father is the point that initial judgments of trustworthiness and confidence are being made. A whiff of stress signals and the father becomes immediately wary. Upon leaving a job interview, the opening of the office door creates just enough wind movement to carry those underarm secretions to the nose of the interviewer and when they sit down to write their initial thoughts, those lingering molecules stimulate the brain to drop those ratings of competency down a few notches. This is the power of aromas to alter our perceptions and decisions. This research shows that the influence of these odors, in the form of neural activity, reaches beyond the emotional centers of our brains and into those rational and cognitive areas of the brain. These odors deftly push our judgments slightly more positive (if the right cologne is used) or more negatively.

Although the research within the area of human chemical ecology is still rare, the possibilities are potentially vast. Imagine a series of colognes where one could choose the right aroma that would evoke the desired behavior in other people. A dose of lavender, vanilla, with a hint of rose water might be the right choice if appearing contrite is necessary. On a different day, a hearty musk with some over tones of fresh rain is the right selection if power and control are the messages to be sent. Instead of a closet of suits, dresses, ties, and scarfs to be selected for the day, one could have an old school apothecary chest with all of the elemental odors that could be mixed together in order to have the cologne of the day. Instead of clothes, these would be invisible cloaks that subtly influence the outcome of a key meeting.

10.7 The Aroma of Love

During one of my class periods on introductory biology, I asked the students to name strong selective forces. That month of the course was dedicated to evolution and we were covering different aspects of natural selection. One of my students, in all seriousness, shouted out that love was the strongest force of all. I was not quite expecting this type of answer, so in an attempt to focus their thought processes on more biologically relevant answers, I quickly replied that love doesn't exist within biology.

The psychological concept of love has very little to do with the detailed mechanics of evolutionary biology, but within the context of human ecology, the student is most likely right. Love is the most powerful positive emotional judgment we can make about individuals in our society. Competence and trustworthiness is certainly an important aspect of one's persona that is critical to control, but deciding to love an individual should be a lifetime commitment. For the purposes of this section, I want to make a distinction between sexual attraction or those types of signals covered in Chap. 8 and the emotion of love. The former entails sexual readiness and physical attraction whereas the emotional concept of love entails trust, companionship, confidence, as well as attraction. These distinctions in concepts lead to differences in the information carried by the chemical signals involved in these choices.

One of the hallmarks of scientific studies is the ability to have control and experimental groups. The best science is done where the two groups are identical except for one factor which should be the factor under examination. Working with humans and the role of chemical signals in our decision making are complex because constructing a control group of humans and removing chemical signals from the equation are exceptionally difficult. We exude chemical signals all of the time, and our unique experiences make it problematic to have true control groups. Luckily, for the research on the smells of love, a control group can be found in anosmics.

Dr. Thomas Hummel, University of Dresden, and his colleagues performed a survey study of men and women who were born without the sense of smell and compared those findings with a similar group of people who had the their olfactory abilities intact. In regard to companionship and mate choice, the inability to smell those subtly important chemical signals had different impact on men and women. For women, the lack of the sense of smell did not impact the number of sexual partners they had throughout their lifetime. Yet, when they were in a relationship, the women, insensitive to the social cues in chemical signals, were a lot more insecure about those relationships. The ability to smell significantly increased the level of confidence and security that women attached to their partners. In addition, the general social insecurity for anosmic women was increased. Anosmac women tended to avoid eating with others and reported increased anxiety about their own body odor. Finally, during social interactions with any other individuals, these women reported problems during those social interactions. The ability to perceive social cues through chemical signals was critical to leading a life of social engagement with reduced anxieties.

Men reported similar effects as the women did. They, too, lacked confidence in social situations, but did not demonstrate the partnership insecurity shown by female participants without the ability to smell. The largest difference between men and women was found in the number of sexual partners. Men missing their sense of smell had significantly fewer sexual partners than their female counterparts. The social insecurity created from missing information gathered through their sense of smell made the men more shy or insecure about approaching women for romantic or sexual interest.

Beyond the aromatic judgments of trustworthiness and competence, this work shows that deep personal connections such as those found in love are profoundly impacted by the chemical signals swirling around us every single minute of every single day. Without the emotional information carried by those molecules, we are less sure of ourselves and our partners. Romantic dinners become a source of stress and confusion as the mind is focused on wondering about body odors, the quality of the food, and the potential fidelity of the partner. The smell of our lover's perfume no longer connects to our emotions. The kiss becomes purely a physical event rather than an opportunity to smell the sweet and endearing cheek of our partner. We may have physical attractions to others using our eyes, but the decision to love is made with the nose. The aroma of love builds an unbreakable bridge of molecules that connects our emotions and cements that relationship.

10.8 The Revealed World of Odors

On a recent trip, I found myself in Phoenix Arizona for an educational conference. I am usually the earliest riser among my traveling companions and after showering and dressing, I partake in what I would call a mini-walkabout. I perform this ritual just about every time I am traveling for business (either science or educational conferences or for invited presentations). This is time that I take to explore the cities, hoping to discover the small off the beaten path restaurant, or odd business that has some interesting wares to sell. In addition, this is my morning foraging run to find a breakfast that is not quite the common staple found in the same old hotel offerings. As I exited the hotel I took a good look at the surrounding buildings so that I could construct a mental map with the spatial relationships of the skyscrapers and then I left in some random direction. At each intersection, I glanced up and down the side streets to see if something piqued my interest.

Early mornings are quite interesting in cities as the streets are not quite at the peak of the morning rush hour, the evening work shift is beginning to shut down, and the shops offering coffee and a quick bite have begun to turn on their lights. Luckily, I was in Phoenix at a perfect time of the year. A light jacket was all that I needed as I exited the hotel lobby and headed off deeper into the heart of the city. I rounded a bend and saw in front of me a television studio with a morning show being televised live. A small crowd of dedicated viewers had located themselves outside of the window in hopes of being filmed. As I passed the small crowd, a light breeze kicked up some of the watchers' perfumes and I got hit with at least four different fragrances. I paused for a second to take in these odors wondering if I could train myself to mentally trace each plume to its owner. I moved on to discover more of this city.

As I approached the corner, I noticed steam rising out of the manhole covers. As I crossed the street, the white plume enveloped me and I found myself processing a very different set of odors. The smell was a combination of sewer smells that must have been beneath the street, the slight hint of fuel oil, and the smell of rotting food. Unlike those women in front of the television morning show, I did not pause to try to trace these plumes to their sources. Unpleasant smells were to be avoided this morning. I did think about the employees whose job was to work within this type of environment. They lived in a completely different odor landscape than the one I had in my campus world. The machinery and operations of the city created a distinctive combination of mechanical smells (oil and gas), food (both new food being produced by restaurants and that food discarded in garbage bins), and the perfume and colognes of the inhabitants. My world was dominated by the far more pleasant smells of the campus greenery when I am outside and the symphony of body odors of 150 plus students sitting in a large classroom.

Heading down the street, I passed a coffee shop and the aromas drew me in for a quick bite. Walking up to the counter, I placed an order for a double espresso. Pure coffee without the extra add-ins of milk, sugar, and flavoring allowed the unaltered aroma of the roasted bean to be sampled by my waiting olfactory system. As the barista started to extract the drink from the ground coffee, my nose was greeted by a mixture of earthy wood smells along with overtones of browned sugar. As I stood at the end of the counter, another early morning patron entered the shop, in search of a similar kick of caffeine that drew me inside. The movement of the door brought a front of air that moved by the pastry display toward my position at the end of the counter. As the air passed by the pastries, molecules of sugar, flour, spices, and flavorings were picked up by the movement of air. The espresso odor mixed with vanilla, chocolate, fresh baked bagels, and of course, my favorite the pumpkin muffin. During my little trip, my odor landscape shifted from the hotel lobby, to the television studio, onto the manhole cover, and finally, to a far more pleasant coffee shop.

These habitats are the sensory landscape under which chemical signals influence my daily interactions. During my field seasons in the summer, I wade through pristine first order streams, jump in bogs, hunt crayfish on the shores of lakes, and dive to dark, cold, and cloudy bottom of small lakes in the middle of the state of Michigan. These habitats are the environments of my favorite olfactory animal, the crayfish. To understand how my animal uses chemical signals to make daily decisions, I need to understand the habitat and the types of signals typically found in that habitat. The walk through Phoenix is just one of many different sensory worlds that humans inhabit. Each of these habitats, whether a large inner city metropolis or in the middle of a farmer's monoculture field, is the sensory landscape for our chemical signals.

Whether we are consciously aware of this landscapes or anosmically wandering around this landscape (akin to being blind in an art gallery), our behaviors are heavily influenced by these signals. We make judgments on the competency of our fellow passengers on planes, we are caught up in the plume of emotion of our partners, we chemically bond to our children even before we name them, we attach motivation to patrons in coffee shops, and we develop a belief in the honesty of our lovers through these signals. This chapter is not really the end point on how odor signals influence and guide our lives. It should serve as a beginning or awakening to a new sense and how we can use that sense in meaningful ways.

10.9 The Endless River of Aroma

We are aromatic creatures. Life by its very nature is chemically oriented, so to be alive means to produce chemicals and exchange those chemicals with the outside world. As such, we leave endless rivers of aromas trailing behind us with every step we take. We choose to ignore these odors and focus primarily on those visual and auditory signals that dominate our thought processes. Just like my colleague's father, who would diagnose his patients based on the smell of their house, it is possible to use these signals in a more fruitful way. In many of the ways covered in this book, we already perform these tasks at a subconscious level. I think it is time that we bring these processes to the forefront of our cognitive thinking processes. As I stated in Chap. 1, my hope was that in reading the many different and disparate animal stories contained in these chapters, the reader has an enhanced perspective of chemical signals in nature. To use Plato's allegory, let's bring our noses (and olfactory brain) out of the cave and expose them to the smells of the world. Maybe those sojourns out into the numerous rivers of aromas would be more pleasant if one achieves this increased awareness.

If we become aware of these rivers of odors around us, maybe we can occasionally take our nose ashore and make more informed judgments in social situations. We can determine the competency of an individual based off a skill set or the bond with our family based off of actions rather than based off a set of sniffs. The subtle perfume or body odor could be just a trick of the mind as the chemicals tug at the emotional centers of the brain. On the other hand, wandering the world of odors more fully aware of the complexity of information around us would seem to make our lives more enjoyable. Strolls through gardens or bakeries would be as delightful and as nuanced as Pink Floyd's *Dark Side of the Moon* or Georgia O'Keefe's flowers as we focus our powers of olfaction on each individual odor as well as the suite of aromas dancing on the wind (Fig. 10.2).

This brings me to a third purpose to writing this book. As a practicing ecologist, I would like to lessen the negative impact that some human activities have on the natural world. Since, as a society, we are fairly unaware of the role that chemical signals play for many different animals, some of our society's activities are harmful to the normal behavior of many different plants and animals. Creating odor pollution can be just as deadly and dangerous as other forms of pollution. For example, exposing some animals to low levels of herbicides doesn't kill them directly but

Fig. 10.2 Endless river of odor

eliminates their sense of smell. Becoming anosmic, these animals can't find food, avoid predators, or even find mates.

As a final challenge to the budding human chemical ecologists, our job should be keeping our aromatic world as clean as possible. Astronomers complain about the fact that light pollution keeps us from seeing the complexity of the Milky Way in the night time skies. Getting the chance to glance up at the night sky in various dark parks or through powerful telescopes is breathtaking. Laying on my back in the middle of a darken meadow during my summer field season, I can see the milky way with my naked eye and the wonder of the galaxy opens up for me. In a similar vein, exploring the smells of a fresh cut lawn, a spring rain, a fall forest, or even a warm pumpkin muffin is equally, and literally, breathtaking. Standing in the middle of an old growth forest as a gentle breeze delivers the aromas of nature to my waiting nose, I am connected, deeply connected, with the natural world. The smoke stacks, car pipes, cigarettes, and heavy cologne is just as damaging to our ability to smell our world as light is to the astronomers. Hopefully these stories have heighten our own awareness of our connection with nature and our place within it by unveiling how intimately connected we are to chemical signals.

Further Readings

Ackerman, Diane. 1991. A Natural History of the Senses. Vintage Press.
Blodgett, Bonnie. 2010. Remembering Smell: A Memoir of Losing—and Discovering—the Primal Sense. Houghton Mifflin Harcout.
Burton, Robert. 1976. The Language of Smell. Routledge & Kegan Paul Press.
Agosta, William. 2000. Thieves, Deceivers, and Killers: Tales of Chemistry in Nature. Princeton University Press.
Agosta, William. 1995. Bombardier Beetles and Fever Trees: A Close-Up Look at Chemical Warfare and Signals in Animals and Plants. Basic Books.
Herz, Rachel. 2008. The Scent of Desire: Discovering Our Enigmatic Sense of Smell. Harper Perennial.
Pert, Candace. 1999. The Molecules of Emotion. The Science Behind Mind-Body Medicine. Simon & Schuster.
Shepard, Gordon M. 2013. Neurogastronomy: How the Brain Creates Flavor and Why It Matters. Columbia University Press.
Turin, Luca. 2007. The Secret of Scent: Adventures in Perfume and the Science of Smell. Harper Perennial.
Sell, Charles S. 2014. Chemistry and the Sense of Smell. Wiley.
Suskind, Patrick (John E. Woods Translator). 2001. Perfume: The Story of a Murderer. Vintage Press.

P.A. Moore, *The Hidden Power of Smell*, DOI 10.1007/978-3-319-15651-4

Index

P.A. Moore, *The Hidden Power of Smell*, DOI 10.1007/978-3-319-15651-4